Studies in Systems, Decision and Control

Volume 404

The series "Studies in Systems, Decision and Control" (SSDC) covers both new developments and advances, as well as the state of the art, in the various areas of broadly perceived systems, decision making and control–quickly, up to date and with a high quality. The intent is to cover the theory, applications, and perspectives on the state of the art and future developments relevant to systems, decision making, control, complex processes and related areas, as embedded in the fields of engineering, computer science, physics, economics, social and life sciences, as well as the paradigms and methodologies behind them. The series contains monographs, textbooks, lecture notes and edited volumes in systems, decision making and control spanning the areas of Cyber-Physical Systems, Autonomous Systems, Sensor Networks, Control Systems, Energy Systems, Automotive Systems, Biological Systems, Vehicular Networking and Connected Vehicles, Aerospace Systems, Automation, Manufacturing, Smart Grids, Nonlinear Systems, Power Systems, Robotics, Social Systems, Economic Systems and other. Of particular value to both the contributors and the readership are the short publication timeframe and the world-wide distribution and exposure which enable both a wide and rapid dissemination of research output.

Indexed by SCOPUS, DBLP, WTI Frankfurt eG, zbMATH, SCImago.

All books published in the series are submitted for consideration in Web of Science.

More information about this series at https://link.springer.com/bookseries/13304

Abdullah Aamir Hayat · Shraddha Chaudhary ·
Riby Abraham Boby · Arun Dayal Udai ·
Sumantra Dutta Roy · Subir Kumar Saha ·
Santanu Chaudhury

Vision Based Identification and Force Control of Industrial Robots

Springer

Abdullah Aamir Hayat
Singapore University of Technology and Design
Singapore, Singapore

Riby Abraham Boby
Innopolis University
Innopolis, Russia

Sumantra Dutta Roy
Department of Electrical Engineering
Indian Institute of Technology Delhi
New Delhi, India

Santanu Chaudhury
Indian Institute of Technology Jodhpur
Jodhpur, Rajasthan, India

Shraddha Chaudhary
Department of Electrical Engineering
Indian Institute of Technology Delhi
New Delhi, Delhi, India

Arun Dayal Udai
Department of Mechanical Engineering
Indian Institute of Technology(ISM)
Ranchi, Jharkhand, India

Subir Kumar Saha
Department of Mechanical Engineering
Indian Institute of Technology Delhi
New Delhi, Delhi, India

ISSN 2198-4182 ISSN 2198-4190 (electronic)
Studies in Systems, Decision and Control
ISBN 978-981-16-6989-7 ISBN 978-981-16-6990-3 (eBook)
https://doi.org/10.1007/978-981-16-6990-3

This Springer imprint is published by the registered company Springer Nature Singapore Pte Ltd.
The registered company address is: 152 Beach Road, #21-01/04 Gateway East, Singapore 189721, Singapore

Dedicated to Teachers and Parents.

Preface

The purpose of this book is to showcase the explored unique interdisciplinary technologies with which a complete functional system is built and demonstrated. Here, a complete functional system refers to the solution developed around precision-based assembly task, namely, "peg-in-tube/hole". To accomplish this task, at hardware level, it utilizes a position-controlled industrial robot, sensors like monocular camera, two-dimensional laser scanner, force–torque sensor and pneumatic grippers, calendrical pellets, metallic bin, workspace where manipulation is to be done, to mention a few. Moreover, the automation of assembly task for accurate pick-up of featureless objects kept in cluttered fashion and its insertion is done using developed algorithms integrated at software level.

The main features of this book and distinguish it from other texts are:

- This book provides a thorough exposure to its readers on the execution of ideas by developing an integrated system for assembly tasks with a conglomeration of multiple technologies.
- The "bin-picking" task presented here is unique with the picking of featureless objects as kept inside a dark bin, unlike a typical Amazon picking challenge.
- How to develop a system for "bin-picking" and methods to combine it with subsequent processing steps to complete the assembly task. Moreover, how the choice of technology, like robots, sensors, etc., can be used to develop an end-to-end technology for an assembly task.
- Transforming a position-controlled industrial manipulator using the identified model of a robot for active/passive force control-based manipulation and safer assembly for "peg-in-tube/hole" task.
- Detailed demonstrable examples with its statistical analysis make understanding and implementation of advanced topics easier.

Industrial manipulator's application for the assembly task has always been a research hotspot. Traditional methods involve pick-up and placement of the object using taught positions and the feedback force control. In this monograph, we focus on the automated and precision assembly task using computer vision-based identification and force control of an industrial robot.

Pose estimation and detection of objects is a quintessential computer vision problem. This monograph proposes a machine vision system to grasp a textureless object from a multi-layered cluttered environment. It contributes to a unique vision-based pose estimation algorithm, collision-free path planning, and a dynamic changeover algorithm for final placement. To achieve it using the vision system, namely, a monocular camera and the two-dimensional laser scanner, data are fused together to obtain an object's accurate pose. The calibration technique was developed to get meaningful data from multiple sensors in a robot frame. Next, the Sensitivity Analysis (SA) of various parameters contributing to the error in pose measurement of the end-effector is carried out. The results from the SA is incorporated for the calibration of a robot that enhanced the pose accuracy.

Identification of kinematic and Dynamic Parameters (DP) of a robot plays a crucial role in model-based control of robotic systems. Kinematic identification deals with finding an accurate functional relationship between the joint space and the Cartesian space, whereas, in dynamics, it is about estimating the unknown DP in the model. The complete information about the Kinematic Parameters (KP) is used in the dynamic identification and control of robots. The modeling for dynamic identification is done using the DeNOC-based approach. This model relates the joint forces/torques and the vector of unknown dynamic parameters using the Regressor Matrix (RM). The RM is a function of KP and the joint trajectory of a robot. Hence, the identified KP from stage one was taken, and for joint trajectory, data were logged from the robot controller in the experiment. In the experiment, the friction and actuator's models are also incorporated for more accurate modeling. The torque reconstruction using the identified model showed a close match with the actual values.

The identified model was used to implement current or torque control at the joints using a position-controlled robot. Hence, the current sensing based on the DC motor controller is discussed in this monograph that can be used in any low-powered robot that intends to do compliant tasks. The force control strategy developed implements a parallel active-passive force controller to perform peg-in-tube assembly tasks. While doing so, we proposed a depth-profile-based search technique for localization and insertion of a pellet into the tube. The assembly task was performed under different orientations of tube, namely, horizontal and vertical tubes. Moreover, various clearances between the tube and pellets ranging from 0.05 to 2.00 mm are considered in experiments. The findings from the experiments verified the effectiveness of the proposed approaches.

Singapore	Abdullah Aamir Hayat
New Delhi, India	Shraddha Chaudhary
Dhanbad, India	Arun Dayal Udai
Innopolis, Russia	Riby Abraham Boby
New Delhi, India	Sumantra Dutta Roy
New Delhi, India	Subir Kumar Saha
Jodhpur, India	Santanu Chaudhury

Acknowledgements

We would like to thank all who have directly or indirectly helped in the preparation of this book. Special thanks are due to the Indian Institute of Technology (IIT) Delhi, where the first four authors completed their PhD thesis. We also thank project staff and researchers who worked under the Programme for Autonomous Robotics and others with whom we had many discussions about life and education that may have influenced the presentation of this book directly or indirectly. Special thanks are also due to our respective family members.

We would like to express our sincere thanks to the Bhaba Atomic Research Centre (BARC) Mumbai for funding the project with Project code RP02346 titled Setting up of a Programme in Autonomous Robotics (2010-2016) spearheaded by Prof. Santanu Chaudhury. The team from BARC headed by distinguish scientist Mr. Manjit Singh, Dr. Debanik Roy, Dr. Deepak Badodkar, and Dr. Prabir K. Pal. Also, the researchers and co-investigators from BARC as Dr. Amaren P. Das, Ms. Varsha T. Shirwalkar, Dr. Kamal Sharma, and Mr. Sanjeev Sharma. The project has inspired the authors to carry out the research related to the topics of this monograph.

We extend our thanks to all the members associated with the project at PAR Lab, IIT Delhi. We also thank Dr. Ravi Prakash Joshi, Dr. Mohd Zubair, Rajeevlochana G. Chittawadigi, Punit Tiwan, Mayank Roy, Tanya Raghuvanshi, Faraz Ahmad, Vinoth Venkatesan who helped in physically implementing some of the tasks mentioned in the study. All authors are most deeply grateful for the superb support given to them by their respective families during writing of this book. The first author wish to acknowledge the National Robotics Programme under its Robotics Enabling Capabilities and Technologies (Funding Agency Project No. 192 25 00051) for the support during his postdoctoral with Dr. Mohan Rajesh Elara at SUTD.

No words can duly represent the appreciation which the authors have for their respective families for giving the authors space and time out of their personal times in order to complete this work.

Finally, we express our sincere gratitude to the reviewers of the book and to the team at Springer, especially Ms. Silky Abhay Sinha and Mr. Lokeshwaran, for helping bring this book to its final form.

Singapore — Abdullah Aamir Hayat
New Delhi, India — Shraddha Chaudhary
Dhanbad, India — Arun Dayal Udai
Innopolis, Russia — Riby Abraham Boby
New Delhi, India — Sumantra Dutta Roy
New Delhi, India — Subir Kumar Saha
Jodhpur, India — Santanu Chaudhury

Units and Notation

The international System of Units (SI) is used throughout this book. The boldface Latin/Greek letters in lower and upper cases denote vectors and matrices, respectively, whereas lightface Latin/Greek letters in lower case with italic font denote scalars. In any deviation in the above definitions, an entity is defined as soon as they appear in the text.

Contents

About the Authors

Dr. Abdullah Aamir Hayat received his Bachelor of Technology in Mechanical Engineering and Master's from the Zakir Hussain College of Engineering and Technology (ZHCET), AMU, India. He completed his Ph.D. in 2018 from the Department of Mechanical Engineering at Indian Institute of Technology (IIT) Delhi. He worked as a Junior Research Fellow (JRF) at the Programme for Autonomous Robotics (PAR) Laboratory under the project granted by BARC/BRNS Mumbai during his Ph.D. He assisted and provided solutions for the research organizations and companies like CMERI Durgapur, Brabo Robotics and Automation, TATA Motors, Pune India, and Oceania Robotics Singapore. He has published several peer reviewed articles, 2 design patents, and reviewed articles for major conferences and journals. His research interests include kinematic identification, calibration, multibody dynamics, design thinking and innovation, and reconfigurable robots. He is working as senior research fellow with the Engineering and Product Development Pillar at Singapore University of Technology and Design (SUTD).

Dr. Shraddha Chaudhary received her Bachelor of Technology in Electronics and Communication Engineering from Bhartiya Vidyapeeth College of Engineering (BVCOE), GGSIPU, India, in 2007 and Master of Technology in 2012 from Delhi Technological University (formerly DCE). She completed her Ph.D. in 2018 from the Department of Electrical Engineering (Computer Technology) at the Indian Institute of Technology (IIT) Delhi. She worked as a Junior Research Fellow (JRF) at the Programme for Autonomous Robotics (PAR) Laboratory under the project "Vision guided control of a Robotic manipulator" 2010–2016) granted by BARC/BRNS Mumbai during her Ph.D. Her research interests include computer vision, robotics, and machine learning. Currently, she is working as a principal project scientist for NOKIA's outsourced project related to mobile robotics at the IIT Delhi.

Dr. Riby Abraham Boby did his Bachelor's from National Institute of Technology (NIT) Karnataka, Master's from Indian Institute of Technology (IIT) Madras and Ph.D. from Indian Institute of Technology (IIT) Delhi. He has been an exchange student at Technical University Munich, Germany and has been involved in academic collaboration with Graduate School of IPS, Waseda University, Japan. His research interests lie in the areas of machine vision, robotics and biometrics. He has been actively involved in research since 2007, and published more than 20 articles in international journals and conferences. He has also been a reviewer for major international conferences and journals in these areas. He is working as Senior Researcher at Innopolis University, Russia from 2019, and is involved in conducting research on calibration of robots using vision systems.

Dr. Arun Dayal Udai is a Faculty in the Department of Mechanical Engineering, Indian Institute of Technology (ISM) Dhanbad, India. He has over 13 years of teaching experience in Robotics, Automation, and CAD domains. He completed his Ph.D. in 2016 from the Indian Institute of Technology (IIT) Delhi, India with a specialization in Force Control of Industrial Robots. He pursued his Master's in Mechanical Engineering Design from BIT Mesra, India in 2006 where he worked on Biped Robots. His research interests include robotics, mechatronics, and CAD. He has published in various peer-reviewed journals and conferences. Currently, he is making 6-RSS Parallel Robots—Aruni and Santulan which are useful for academics and research purposes and provides consultancy to various industries in robotics and automation.

Prof. Sumantra Dutta Roy is a B.E. (Computer Engineering) from D.I.T., Delhi (1993), and completed his M.Tech. and Ph.D. degrees at the Department of Computer Science and Engineering, Indian Institute of Technology (IIT) Delhi, in 1995 and 2001, respectively. He started his career teaching and research in the Department of Electrical Engineering at IIT Bombay. He is currently Professor in the Department of Electrical Engineering at IIT Delhi. He is a recipient of 2004 INAE Young Engineer Award (Indian National Academy of Engineering), and the 2004–2005 BOYSCAST Fellowship of the Department of Science and Technology, Government of India. His research interests are in computer vision and image analysis, video and image coding, biometrics, music information retrieval, and medical informatics.

Prof. Subir Kumar Saha a 1983 mechanical engineering graduate from the RE College (now NIT), India, completed his M.Tech. from Indian Institute of Technology (IIT) Kharagpur, India, and Ph.D. from McGill University, Canada. Upon completion of his Ph.D., he joined Toshiba Corporation's R&D Centre in Japan. After 4-years of work experience in Japan, he has been with IIT Delhi since 1996. He is actively engaged in teaching, research, and technology, and completed projects worth more than USD 1.0 million. He established the Mechatonics Laboratory at IIT Delhi in 2001. As recognition of his international contributions, Prof. Saha was awarded the Humboldt Fellowship in 1999 by the AvH Foundation, Germany, and the Naren Gupta Chair Professorship at IIT Delhi in 2010. He has also been a visiting faculty to universities in Canada, Australia, and Italy. He has more than 175 research publications in reputed journals/conference proceedings and delivered more than 150 invited/keynote lectures in India and abroad.

Prof. Santanu Chaudhury is the Director of Indian Institute of Technology (IIT) Jodhpur. Prof. Chaudhury holds B.Tech. (Electronics and Electrical Communication Engineering) and Ph.D. (Computer Science & Engineering) degrees from IIT Kharagpur. He joined as Faculty Member in the Department of Electrical Engineering, IIT Delhi in 1992. He has served as Director of CSIR-CEERI, Pilani, during 2016–2018. Prof. Chaudhury is a recipient of the Distinguished Alumnus award from IIT Kharagpur. He is a Fellow of Indian National Academy of Engineers (INAE) and National Academy of Sciences (NAS). He is a Fellow of International Association Pattern Recognition (IAPR). He was awarded the INSA (Indian National Science Academy) Medal for Young Scientists in 1993. He received ACCS-CDAC award for his research contributions in 2012. A keen researcher and a thorough academic, Prof. Chaudhury has about 300 publications in peer reviewed journals and conference proceedings, 15 patents and 4 authored/edited books to his credit.

Chapter 1
Introduction

The potential use of robots has expanded manifolds. Since the first modern industrial robot in 1960s, robots are finding applications in various manufacturing sectors, warehouse management, telesurgery, rehabilitative task, hazardous environment, maneuvering, and surveillance in an unknown environment. The International Federation of Robotics defines an industrial robot as "automatically controlled, reprogrammable, multipurpose manipulators programmable in three or more axes". Typical industrial robots are large and bulky automated machines. They are also inexpensive without any integrated or embedded sensors, like a joint torque sensor that makes a robot with embedded sensor-sensitive, safe, and expensive. Industrial robots have good repeatability but poor accuracy, and are position controlled brute machines that apply full force to reach the commanded pose, i.e., position and orientation. It results in several challenges during a manipulation and assembly task and it becomes more challenging during a precision based assembly where the mating objects have negligible clearance.

The robotic assembly task includes the pick-up and placement of an object at a dedicated place. There are several applications for robotic assembly operations like drilling, bolt insertion, etc. It not only includes parts that are larger in a dimension like gearboxes, piston, and carton box but also small objects like electronic components and pellets. An essential and widely observed assembly task can be categorized under "bin-picking" for the pick-up of objects and "peg-in-hole" for feeding picked objects. Moreover, a bin-picking task using robots requires gripping objects placed in a random manner, occluded, and even in several layers on a surface or bin. Also, the peg-in-hole assembly task requires the insertion of picked objects inside a hollow

---

Supplementary Information The online version contains supplementary material available at https://doi.org/10.1007/978-981-16-6990-3_1.

A. A. Hayat et al., *Vision Based Identification and Force Control of Industrial Robots*,
Studies in Systems, Decision and Control 404,
https://doi.org/10.1007/978-981-16-6990-3_1

region, and if that is of a tube, it is labeled as "peg-in-tube". This work focused on developing methods and algorithms for performing safely the bin-picking task of metallic cylindrical pellets kept randomly in layers inside a metallic bin that results in unidentified features while viewing from outside. The pellets have to be picked using a six-degrees of freedom industrial robot (without any embedded joint torque sensors). Further, the picked pellets need to be inserted inside a tube with limited clearance. The task has to be performed safely, i.e., without damaging the pellets, tube, or environment due to any form of collision.

As discussed above, the precision-based assembly tasks can be accomplished using advances in the vision, sensing methods, modeling, identification, control algorithms, programming, and computational power. The fundamental of robot motion lies in kinematics and dynamics of the robotic manipulator. Unlike humans and living organisms, robotic manipulators explicitly need the information of their geometric or kinematic parameters and dynamic parameters for manipulation. Especially with the increasing complexity of the task that requires high accuracy and interaction of the robot with the environment, precise knowledge of the system, i.e., of a robot and sensors modeled parameters, is required.

Kinematic modeling aims at building a mathematical model of the robot architecture understudy that determines the position, velocity, and acceleration of each link as a function of its geometric parameters. In order to study these in Euclidean space, the links are assigned with frames and a sufficient number of parameters to describe each frame with respect to another. Generally, six parameters are required to completely describe the pose, i.e., position and orientation, of a frame with respect to another one. However, the use of Denvait Hartenberg (DH) parameters simplifies such representation, requiring only four parameters. The architecture of robots can be represented using the DH parameters and these are not explicitly revealed by most of the robot manufacturers. For accurate manipulation, the identification of the DH parameters of an installed manipulator is essential. Moreover, object detection with pose estimation, i.e., position and orientation, is a quintessential computer vision problem. The use of vision-based manipulation is finding application in multiple domains, like, assembly, bin-picking, etc. Figure 1.1 highlights the issues[1] during the vision-based manipulation where it is essential to estimate the accurate pose of an object. Pose estimation of the object requires camera calibration and pickup of these objects using robot requires accurate kinematic calibration, whereas the calibration between the camera and the manipulator is referred to as eye-in-hand calibration.

Sensitivity analysis and uncertainty modeling play an essential role in evaluating robotic pose estimation algorithms, particularly during sensor fusion. The study of local sensitivity and the global sensitivity of the model for sensor calibration is vital. In this context, sensor fusion-based measurements represent a contactless technique that can obtain data from an image-recorded scene using digital cameras and depth-values recorded with a laser scanner in the same frame. To devise the best pose estimation algorithm, one must understand how the uncertainty due to random perturbation affects the input images obtained using a camera, and depth-maps obtained using laser scans and how it propagates through various algorithmic

[1] Question marks inside Fig. 1.1 indicates that the state is unknown or has to be estimated

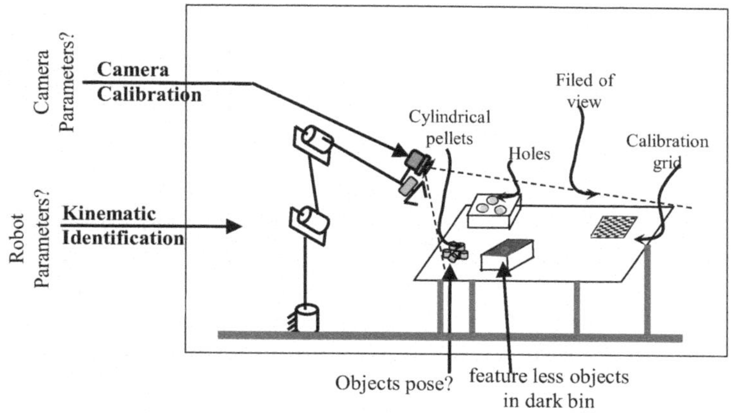

Fig. 1.1 Vision-based identification of an object pose and kinematic identification of a manipulator

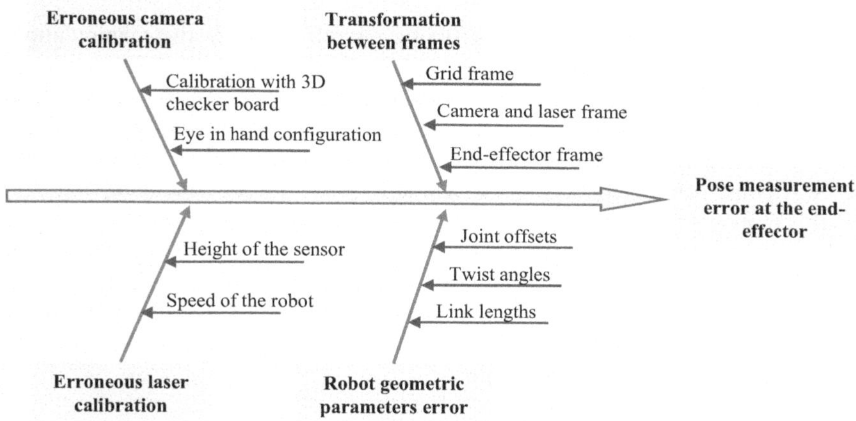

Fig. 1.2 Various parameters effecting the measurement of the pose

steps resulting in an error on the final estimates. Figure 1.2 depicts the various sources from where the error in the pose measurement is caused. This highlights the fact that accurate kinematic parameters and transformation matrices estimation is essential.

Accurate pose estimation is not enough for an industrial manipulator to complete assembly tasks. Observing one of the fundamental characteristics of any living being is its disposition or tendency to yield to others' will. This property of a human is phenomenal in terms of the motion capabilities of its limbs. On the other hand, industrial robots are brute machines that can only be programmed for its end-effector's pose or joint angles. Once commanded, it applies its full potential to execute the desired command without considering any obstacles on its way. It poses a serious problem with any industrial robot without dedicated joint force/torque sensors, as it can cause severe damage to the robot or its environment. With the increasing

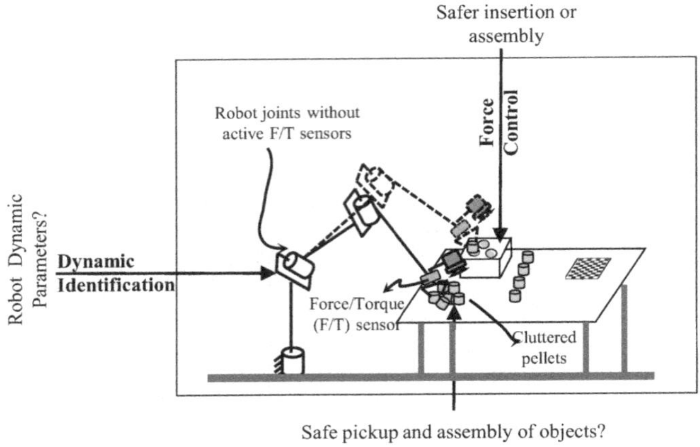

Fig. 1.3 Force-based assembly of objects using a manipulator

demand for industrial robots to perform robotic assembly, cooperative manipulation, end-effector, or link-level interactions with the environment, pure position control strategy becomes inadequate. Hence, manipulation based on force control is needed for safer assembly tasks, i.e., without damaging the robot, object, and surroundings. Figure 1.3 highlights the issues during the assembly task where the estimation of the robot's dynamic parameters becomes essential for the control of the manipulator.

In this monograph, an attempt has been made to present a framework for vision-based force control of an industrial robot. The "Vision-based" refers to the notion that for any assembly task it requires accurate estimation of the pose, i.e., position and orientation of an object kept within the field of view of the vision sensor or camera. "Force control" is essential for the safe pick and placement or assembly task performed by the industrial manipulator.

1.1 Focus of the Monograph

This monograph provides several new insights into robot-based pick-and-place and assembly tasks as well as calibration of robotic systems. The objective of this monograph is two-fold. First being the vision-based pick and place, i.e., manipulating objects in a cluttered environment using a robotic manipulator discussed in Sect. 1.1.1. Secondly, the problem of assembly in an industrial scenario is closely related to the introduction of sensors on the robotic arm for automation. It has long been recognized that automated assembly lines serviced by robots is one of the most delicate and difficult tasks by industrial robotics. Such tasks, e.g., insertion of a solid cylindrical object called pellets in a hollow cylindrical hole Fig. 1.3, involve two problems, namely, detection of the pose of the object and detection of the hole position. The most tedious phase of the assembly process is mating parts. Force feedback control is applied to achieve the task is highlighted in Sect. 1.1.2.

1.1.1 Computer Vision

Vision is an essential requirement in addressing picking and manipulating objects in a cluttered environment using a robotic manipulator. This is perhaps the most important sensory mechanism for humans as well. The inputs from cameras guide the manipulator and the gripper to perform the task at hand. The motivation is to have an autonomous system to perform the task. It involves the synergistic combination of the two crucial but practical fields of computer vision and robotics. Computer vision is used first to identify the object of interest and then estimate its pose. Subsequently, the robot manipulator performs the task of grasping. The problem is confounded if the objects to be picked are small and without visual features that allow the use of vision techniques. In engineering applications and industries, such objects are very common.

After picking the object, say, a pellet, the pellet will be placed into a hollow cylindrical tube whose exact location may not be known. In such a situation, visual identification of the target location and its pose is needed to guide/enable the manipulator to pick up the pellet by its gripper and then insert it into the tube. This task requires excellent precision and accuracy. Vision is a natural aid in such a synergistic task with its limitation of positioning and orientation accuracy depending upon several factors like camera used, lighting conditions, etc. Even in an imaginary case where the hole position is accurately known, errors in positioning the robot makes it impossible to complete the assembly task. This is why force control becomes essential, as discussed in the next section.

In case of the multi-layered cluttered configuration shown in Fig. 1.4, depth information also plays an essential role in determining the pose of the objects for manipulation. A camera cannot alone capture depth information. Therefore, the information from visual-band sensors 2D camera and laser scanner are fused to obtain pose of the object.

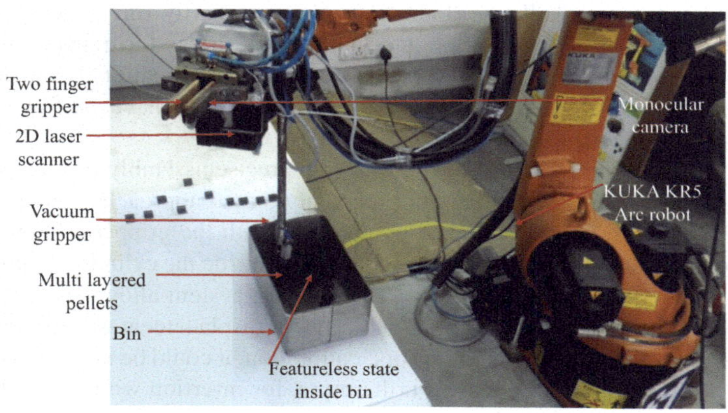

Fig. 1.4 Industrial robot equipped with sensors and grippers at the end-effector

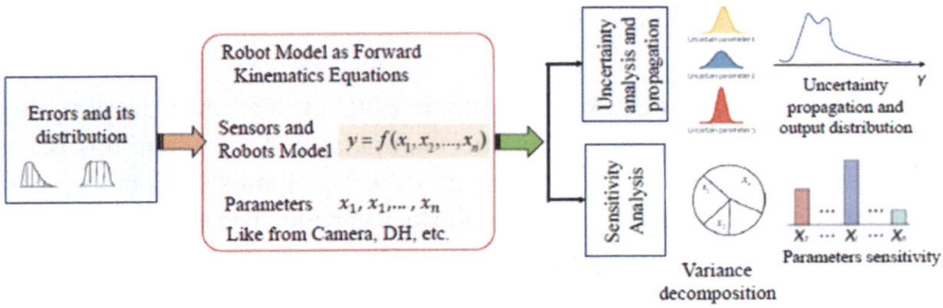

Fig. 1.5 Uncertainity and sensitivity analysis related to robot and camera parameters

Therefore fusing sensors information plays an important role in increasing the efficiency of the system. Hence, it becomes critical to analyze the uncertainity propagated through the sensors reading and also the sensitivity of the robot kinematic parameters. The aspects of uncertainity and sensitivity analysis for a system model and its parameters, namely, robot and sensors muonted on the robot are shown in Fig. 1.5.

1.1.2 Force Control

Industrial robots with human-friendly workspace are most sought after in the current industrial environments where the product life cycle is short spanned. The robot workspace now needs to be frequently accessible for maintenance, measurement, calibration, programming, etc. As a result, the human–robot coexistence becomes essential. Figure 1.6 shows the human obstacle during the path followed by the robot which stops after a safe threshold force value. With an increasing number of collaborative robot tasks with overlapping workspaces in the industry, a robot–robot safe co-existence is a challenge. Industrial robots primarily run on joint position servo actuators with high gear ratio and toothed-belt drives for transmission. The controller can be commanded only for end-effector or joint position. For making the robot compliant or safe to the environment, it is essential to have the force control techniques implemented in the robot.

The application of an industrial robot for the precise assembly task is shown in Fig. 1.6. Here, an object identified after the vision-based approach as mentioned in Sect. 1.1.1 needs to be safely picked up and held with the gripper without hitting the environment, and then the insertion of the pellet inside the cylindrical tube needs to be carried out. Simple fixturing or a crude vision system allows us to align the z-axis position within 0.6mm and rotational angular displacement within a degree or two. Moreover, with the robot and sensors' calibration, it could be improved up to an accuracy within a millimeter scale. Furthermore, for insertion with tight tolerance, the exact hole center needs to be located. This task requires a search approach inspired by human "*where a human can insert a key inside a lock in a completely dark room*".

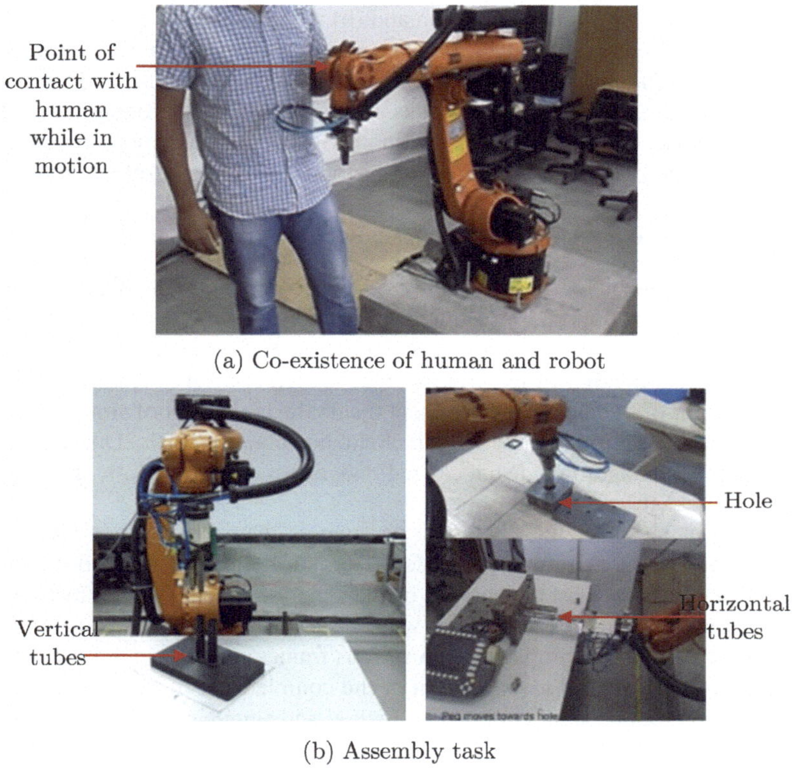

Fig. 1.6 Robot compliance and assembly task

For this approach to be imitated by an industrial robot, as shown in Fig. 1.6b, the dynamic model of the system is essential along with the force/torque (F/T) sensor integrated with the robot.

In this monograph, we present a solution to automate the whole process of bin-picking and assembly task using vision-based force control technique. Also some of the essential codes for integration, calibration, force control, etc., are provided in the Appendices.

1.2 Salient Features of the Book

The vision-based identification and force control of industrial robots reflect on the research and development efforts towards precision assembly tasks by picking featureless (from a vision perspective) objects. The phrase "vision-based identification" covers two broader aspects, namely, (a) estimation of pose and feature of an object say, pellets and hole using sensor fusion and (b) identification of kinematic parameters of a serial robot, its repeatability using a monocular camera. The information on the pose and kinematic parameters of a robot using a vision-based identification technique helps during assembly tasks and accurate offline programming of the robot. The phrase, i.e., "force control of industrial robot" points to converting the position-controlled industrial robot into active/passive force control mode for safer assembly task.

The important contributions in the book are outlined below:

- Calibration approach for different sensors mounted on robots end-effector, namely, 2D laser scanner robots end-effector, and monocular camera in order to utilize the information from multiple sensors in a single frame of reference.
- Method for bin-picking of featureless objects. Here, object is featureless due to the dark color of the metallic object kept inside the bin or box of similar color.
- Using a low-resolution scan with point cloud data obtained with 2D laser scanner and robot movement to guide a localized image search for optimally utilizing the information from multiple sensors
- Sensitivity analysis of the kinematic parameters, namely, Denavit–Hartenberg (DH) parameters are presented in order to assign the quantitative index to these parameters on the basis of their effects on the variation of robots end-effector poses
- Analysis of uncertainties associated with the robot and sensors to quantify the expected errors and its propagation in various frames.
- Monocular vision-based identification of the complete set of kinematic parameters for serial robots using circle point method and singular value decomposition technique.
- Dynamic modeling using the concept of the DeNOC matrices was adopted for the identification of dynamic parameters, which eventually helps in the model-based control of the robot.
- Force controller scheme for existing joint-position controlled robots in the industry so that it can perform active force control tasks using its end-effector, apart from making the robots joints compliant to make it safe against any unpredicted link collision.
- Localization algorithm using force control to perform a peg-in-task assembly task in two orientations, namely, vertical and horizontal using a position-controlled industrial robot.
- Experimental validation of the assembly task by pick-up of featureless pellets kept inside the bin and then inserting into the tube with negligible clearance.

1.3 Layout of the Book

The challenges mentioned in the previous sections and the issues in current literature discussed in the respective chapters motivate us to develop a generic platform-independent framework. We propose a solution for each sub-problem. This mono-

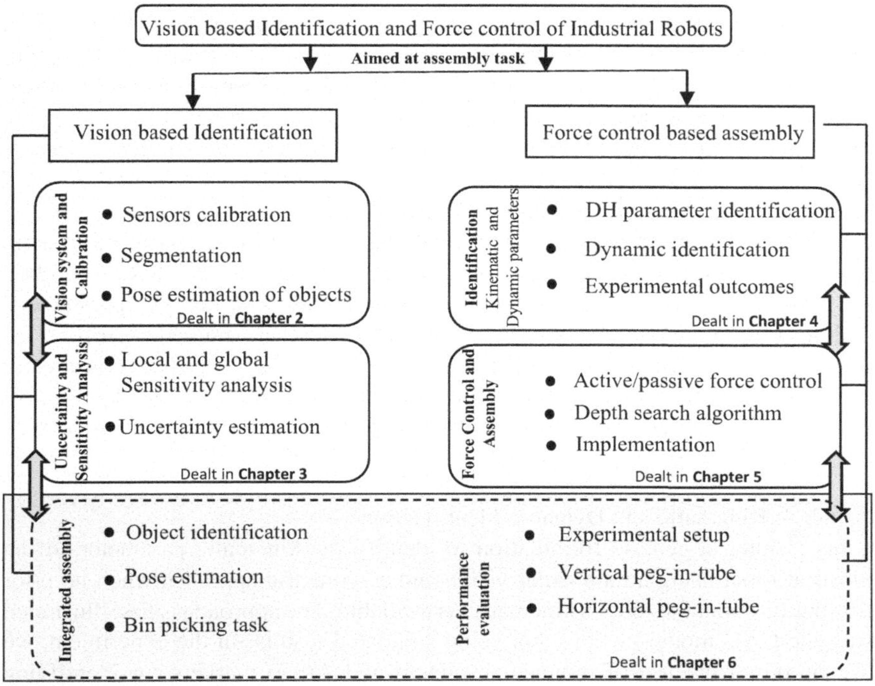

Fig. 1.7 Outline of the topics covered

graph is an amalgamation of vision, robotics, kinematics, calibration, sensitivity analysis, and force-control, to build a complete solution to the object-picking and assembly problem in an industrial scenario. Not only this, to make the research versatile and robust, experiments have been performed using a 6-DOF (KUKA KR5 Arc) robotic manipulator with visual feedback to close the feedback loop.

As shown in Fig. 1.7 this monograph contains six chapters and three appendices. They are organized as follows:

Chapter 1: Introduction
This chapter discusses in brief the scope, objectives, layout, and contribution of the monograph. Challenges involved in the object detection, manipulation, assembly using an industrial robotic arm and practical aspects to these sub-topics are also discussed.

Chapter 2: Vision Systems
In this chapter, we discuss in detail about the calibration of sensors. Need of sensor calibration is also described in this chapter along with the transformation from the base of one sensor to another sensor. This chapter plays vital role in the understand-

ing of concepts dealt in the following chapters. Estimation of the 6D pose of an object in a cluttered environment is dealt in this chapter. Deterministic approach to multi-sensor-based segmentation techniques is discussed. Pose estimation is a difficult problem when it has to be done for a featureless and textureless object. Keeping this in mind, solutions have been proposed using information from multiple sensors.

Chapter 3: Uncertainty and Sensitivity Analysis
This chapter discusses the uncertainty modeling of an information fused system, which can support a bin-picking and assembling task using a KUKA KR5 Arc robot. The sensitivity analysis of the robot's kinematic parameters, i.e., DH parameters, are derived to figure out the contribution of each individual parameters error on the pose of the robot's end-effector in its workspace. Two sensor types are considered: a vision-based sensor for 2D information and a laser scanner to get the depth information. The developed sensor-fused system is applied to plan and optimize the pick and place process for cylindrical objects placed in a clutter.

Chapter 4: Kinematic and Dynamic Identification
In this chapter, a generic formulation to identify the kinematic parameters of an industrial robot using a monocular vision and a geometric approach when no prior information about the robot's kinematics is available. The approach is first illustrated using the CAD model of the robot using simulated results. In the experiment, we have proposed the use of monocular camera and Aruco markers for identifying kinematic parameters and the results were compared with the other measurement devices, namely, total station and laser tracker.

A proper dynamic model of the robot is crucial as it should be able to emulate the real robot, particularly, if the robot moves at relatively faster speeds (typically over 1 m/s). This is done here using the concept of Decoupled Natural Orthogonal Complement (DeNOC) matrices, which is known to offer a recursive forward dynamic algorithm that is not only efficient but also numerically stable. It is presented in this chapter with the experimental results for the industrial robot KUKA KR5 Arc.

Chapter 5: Force Control and Assembly
This chapter proposes an active force control of the end-effector using a wrist force/torque sensor through an external force-control loop and a steady-state error compensator. Further, the existing robots in the industry are attached with an end-effector force–torque sensor to perform any compliant manipulation tasks. Any inaccuracy in these robots that leads to any link colliding with the environment may prove to be disastrous as the joints are run by stiff position controllers. In order to tackle such an event, this work discusses the parallel active–passive force control scheme that can perform active force control at the end-effector and in parallel it is also capable to dynamically adjust the maximum current or torque that any joint can deliver. In parallel, the passive joint compliance was achieved by limiting the currents to

joint motors based on an identified model of the robot under study. The proposed method was implemented on a KUKA KR5 Arc industrial robot and tested for passive compliance. The experimental outcome of the force control algorithms during the assembly task using an industrial manipulator is discussed. The experimental setup for the peg-in-tube task carried out using an industrial robot KUKA KR5 ARC. The vision-based force control approach is used based on the formulations and outcome of the previous chapters.

Chapter 6: Integrated Assembly and Performance Evaluation
This chapter focuses on the integration of vision-based pickup and force control-based assembly tasks. Here we elucidated the assembly task with the pellet insertion inside a hollow tube with significantly less clearances, i.e., the problem is generalized under peg-in-tube. The algorithms discussed in the previous chapters were successfully implemented for the two orientations of the tube, i.e., one in parallel to the direction of the gravity and another in which it is kept horizontal. The statistical results were also presented to evaluate their performance of the assembly task.
References

Appendix A: Vision System and Uncertainty Analysis

In this appendix, the DH notation adopted here, were mentioned along with the concept of dual vectors. The sensitivity index along with the illustrations are given. The technique used to find the pose using the monocular camera is mentioned.
Appendix B: Kinematics, Dynamics, and Control

In this appendix, the expression for the dynamic model of a two link serial manipulator is given. The torque due to gravity was derived for a serial chain robotic system. Again, it utilized the intermediate matrices of forward kinematics to derive the gravity torque of a 6-DoF robotic arm.

Appendix C: Codes Snippets

In this appendix, the sample codes and snippets of the codes are provided related to the calibration, identification, and control schemes. Note that the sample code is shared to give the implementation aspects of the integration task for the vision and force control-based assembly and these are subjected to change as per the programming environment.

1.4 Summary

This chapter of the monograph presented the introduction to the object detection, manipulation, and assembly problem. The chapter highlights aspects in the automation of the pick-and-place using information fusion in industrial robots. Finally, the important features and layout of this book is presented.

Please scan above for videos and refer Appendix C.3 for files detail

Chapter 2
Vision System and Calibration

This monograph highlights how different technologies, namely, vision, robotic manipulation, sensitivity and uncertainty analysis, kinematic and dynamic identification, force control etc., can be combined to get an end-to-end fully functional system. This chapter acts as a window to this treatment to explain perceiving the external environment using sensors. The environment to be sensed contains a large number of symmetrical and identical feature-less, texture-less objects (black cylindrical pellets) randomly piled up in the bin, with arbitrary orientations and heavy occlusion. The task at hand is to use a set of sensors with complementary properties (a camera and a range sensor) for pose estimation in heavy occlusion, accordingly orienting the manipulator gripper to pick up a suitable pellet, and one-by-one. Furthermore, the manipulator avoids collision with the bin walls and identifies and characterizes cases when the object is present in a blind spot for the manipulator. Thus, this chapter lays the foundation for the subsequent chapter, i.e., uncertainty and sensitivity analysis of the vision-based system. These chapters also prove that building a fully functional robust pipeline in real-world scenarios is quite challenging. Particular aspects that are of paramount importance for the successful accomplishment of the task is also presented. Another critical aspect to emphasize here is that it deals with real-world industrial problems, i.e., bin-picking. Different approaches that work best due to variations in assumptions and experimental protocols, e.g., sensors, lighting, robot arms, grippers, and objects, are dealt with in detail. A layout of the literature survey is provided as visual chart to give the major technologies and components involved in this chapter. The contents of this chapter will be suitable for practitioners, researchers, or even novices in robotics to gain insight into real-world problems.

Supplementary Information The online version contains supplementary material available at https://doi.org/10.1007/978-981-16-6990-3_2.

A. A. Hayat et al., *Vision Based Identification and Force Control of Industrial Robots*,
Studies in Systems, Decision and Control 404,
https://doi.org/10.1007/978-981-16-6990-3_2

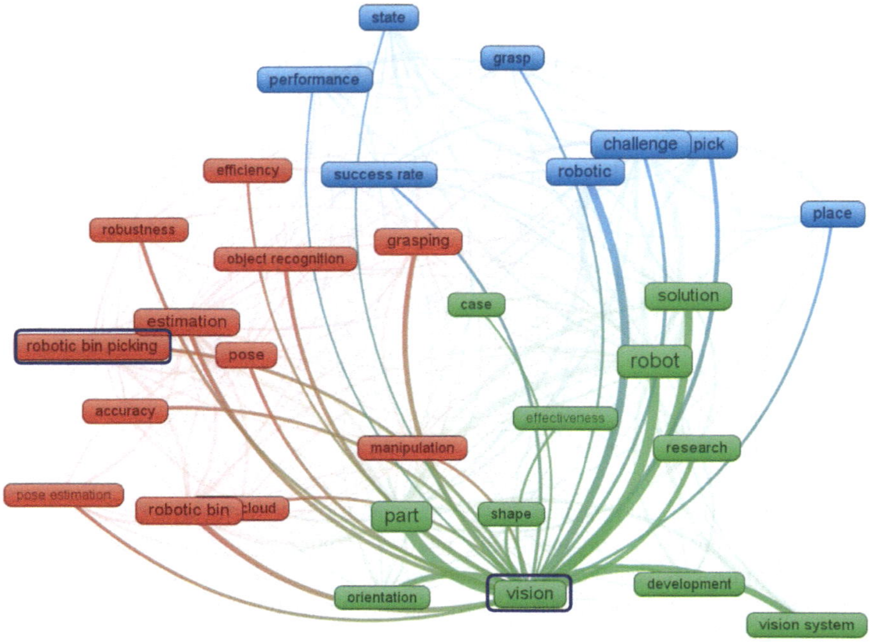

Fig. 2.1 Vision and other components associated with the bin picking task

2.1 Background

Calibration of the sensors mounted on the robot end-effector is the first step before using the complete system. Later the suitable algorithm may be used to process and segment the data from the 2D sensor and camera. The practical aspects involved in implementing the setup is presented. Figure 2.1 shows the components associated with the vision based robotic bin-picking task after the survey of over 150 literature published till the year 2021. The existing approaches are presented next.

2.1.1 Vision Based Solution for Bin Picking

To solve this problem by using vision initially was based on modeling. One approach of solving the problem is based on modeling parts with invariant shape descriptors assuming 2D surface representation [1]. There were other system as well ,which recognized ,scene objects from range data using various volumetric primitives such as cylinders [2, 3].

A flexible bin picking system should address three problems, namely, scene interpretation, object recognition and pose estimation. These approaches mainly dealt with the planar objects and each approach had its benefits and limitations. Approaches like spatial orientation cannot be robustly estimated for objects that are free-from contours. In order to address this limitation most bin picking problem system attempt to recognize the scene objects and estimate their spatial orientation using the 3D information [4]. One of the milestone approach include the use of 3D local descriptors [5] and visual learning methods [6, 7].

Another aspect to bin picking problem is to deal with shape matching of the object so that it could be generalized for any object type. There are number of authors who have proposed shape representation and similarity measure, which are invariant to object deformations [8]. They not only handle intra-class shape variation but also achieve good performance for object recognition. But their requirement for clean segmentation acts as their drawback and hence makes them unsuitable for many application in which the objects are cluttered or overlapped in a dense manner.

To overcome with the drawback of the approach [9] proposed an approach in which shape matching problem is posed as finding the optimal correspondences between feature point, which lead this to an quadratic problem. Reference [10] proposed a contour segment network framework in which shape matching is formulated as finding paths in the network that are similar to model outlines. Ferrari et al. [11] gave another extension to his previous work in which he took edges as k-connected nearly straight contours fragments in edge map. These descriptors which are invariant are further utilized in shape matching framework using voting scheme on a hough space [12]. The 3D object detection from RGB-D data in both indoor and outdoor scenes using machine learning based network named as pointnet [13]. PointNet is highly efficient and empirically, it was shown strong performance at par or even better than state of the art [14].

2.1.2 Learning Based Approaches for Bin Picking Task

The bin-picking tasks generally follows three steps, namely, object localization, recognition, pose estimation by matching the model of object, and planning the grasp action. These three tasks can also be potentially carried out using machine learning techniques as discussed below with its limitations. Object localization/segmentation can be conducted by traditional template matching based on designed descriptors (e.g., scale-invariant feature transform (SIFT) [15, 16]) or deep convolutional neural networks (e.g., mask region-based CNN (R-CNN) [17], Single shot multibox detector (SSD) [18], You Only Look Once (YOLO) [19], or SegNet [20] for 2D object segmentation, and PointNet [21] or PointCNN [22] for 3D object segmentation).

Once we get the segmented points from the first step, object pose can be estimated by cloud point registration, which searches a spatial transform to align two cloud sets [23]. Drost et al. [24] proposed Point Pair Features (PPF) for building a hash table as object model global descriptors and recovering 6D poses from a depth image by a voting scheme, the performance of which under noise and occlusion was improved in [23, 25, 26]. Recently, deep convolutional neural networks that jointly output the semantic segmentation and object 6D pose have been proposed [27–29]. These methods integrate object segmentation and pose estimation, and outperform conventional methods in cluttered scenes when objects have features that are discernible but not for the objects that lack any feature or texture.

Once the object pose is obtained and the object is known, a template-based method can be used to find corresponding grasp points predefined in a database if objects are already exist in the database but computational complexity and cost increases if the objects are novel and not present in the database. In order to pick/grasp an object one need to quantifiable information on certain parameters such as object geometry, physics models, and force analytics [30, 31]. It might be unavailable or difficult to design for bin picking in cluttered scenes. Hence the problem to determine the 6D pose of a texture-less and feature-less object in cluttered configuration is complex and an approach to solve it is discussed in this chapter.

2.1.3 RANSAC Based Algorithms for Bin Picking

Segmentation based methods, namely, 3D RANSAC was implemented in [32, 33], tends to be computationally expensive due to their iterative nature even for small objects whose cloud points are iterated. Optimised RANSAC as in [34, 35] is typically faster but their implementation shows, high false negatives. The localization technique used here computes a local geometric characterstics, but later goes on to implement RANSAC. We were of the opinion of using the generated information and concluding the object recognition on geometrical basis. Looking for geometric approaches in nature, we found [36] uses projection of normals on a Gaussian sphere is faster than RANSAC, but not robust in case of heavy clutter and multiple primitives. This led us to develop a new algorithm which does not use RANSAC based iterative technique to localize the object of concern hence its more robust in the heavily cluttered environment and faster as compared to the already available solutions.

2.1.4 Sensors for Bin Picking

In case of non-planar objects, there occurs a need to incorporate some more information which may be by using some form of a range data. Three dimensional part orientation (3DPO) SYSTEM is a system developed by [37] for recognizing and locating 3D objects from the range data. This system is a two-part recognition model. First, a low-level analysis of range data is carried out followed by a recognition scheme based on model matching. Edges from the range data are recognized by the rst component (curvature is computed from derivatives of up to second order) of the recognition system and classified into circular arcs and straight lines, and the resulting edges are partitioned and classified in a multi-step process. A circular edge is, therefore, has one planar surface and one cylindrical surface adjacent to it but there may be an intersection of two planes for a straight edge. Once, the edge classification process is performed, the low-level analysis indexes all the visible surfaces that are adjacent to the edge features. An unknown object is recognized by the second component of the system by searching the model database for features matching with the features of the object to be recognized. The proposed system works well for a very few models for complex scenes containing clutter and occlusion. Also,the system is better suited for recognition of curved objects.

In different range image applications many further methods belong to 2D template matching [38–40]. This work is based on sensor of fusion, here both camera and laser range scanner is used to localize the object of interest. Data structures for 3D surfaces arise from moving 2D sensors [41] but also from 3D distance sensors. Most applications use line sensors [42], which are moved on external axes 2.5D data is characterized by a 2D-array of distance values, i.e. every pixel represents a value for the distance between the sensor and the object. This kind of data acquisition is getting more attractive in industrial application because of the growing number of range sensors in the market. Range sensors should not be mixed with real sensors delivering real three dimensional (3D) data as they are very sparse and also very expensive. Therefore, only a very small number of applications can take advantage of them.

An attempt was made by [43] to determine the location and orientation of general 3D objects with an approach based on the 3D Hough transform. The effects of translation, rotation and scaling on the 3D Hough space are discussed by the authors. They implemented a hierarchical algorithm to computation of object distortion parameter because of the separable nature of the translation and rotation effects. The developed system showed reliable results only for objects with a simple configuration in scenes containing a single object [44], proposed an object representation scheme that incorporates three concepts such as feature sphere, local feature set and multi attribute hash table. The scheme is used to reduce the complexity of scene-to-model hypothesis generation and verification. This system performs well, just like the previous systems only while dealing with a single-object scene and gave unsatisfactory results for scenes containing a large number of objects. Sundermeyer et al. [45] recently proposed implicit 3D orientation learning for 6D object detection.

A framework was proposed by [6] for simultaneous recognition of multiple objects in scenes containing clutter and occlusion. Matching surfaces by matching points using the spin image representation is the basis of the approach. The spin image is a 3D shape descriptor which depends on the object surface curvature. Thus, being associated with the 3D information, the approach requires a precise range sensor, in order to be efficient. Reliability is best shown when the objects of interest present complex 3D shapes.

2.1.5 *Some Recent Trends*

Enabling robots to manipulate and perceive in a workspaces has years of ongoing research. Despite this long-standing research effort, there continues to be a vast ability gap between the tasks robots and humans are able to accomplish. This fact is perhaps most evident in the various robotic challenges within recent years, like the Amazon Picking Challenge [49], the DARPA Autonomous Robotic Manipulation (ARM) challenge [4] and the Robot Grasping and Manipulation Competition 2016 [50], in which the robots demonstrate their tiny dexterities to achieve certain tasks to help humans in their day-to-day businesses/chores. Amazon Picking Challenge being the latest and most related to the contents of this chapter. In this challenge main task is to pick-and place or robotic manipulation using several types of sensors and actuators [49]. The main challenge lies in the wide variety of products to be packed with various sizes, shapes, weights and degree of fragility. The main task of those problem was to safely plan and execute a grasping action for such objects. Most of the current solutions [51–54] of the robotic pick-and-place task is done by understanding the object model, which normally follows these three steps: object recognition, pose estimation by matching the model of object, and planning the grasp action base on deep learning based approaches. But such methods based on learning based algorithms tend to fail when the object be pick-and-place are texture-less and feature-less. Table 2.1 lists the features of the proposed approach over the other recent methods.

2.2 Calibration

Robotic system requires multiple sensors to be used for the purpose of measuring the external environment like cameras, 3D sensors etc. Calibration of these sensors are briefly discussed in this section of the chapter. Two types of calibration are necessary. One is intrinsic calibration which is determination of the internal parameters of the sensor. The other is extrinsic calibration which deals with the relative pose of the sensor coordinate frame with respect to the robot coordinate frame. Both are explained.

2.2.1 *Setup*

Multiple instances of using camera for picking up objects are available in the literature [55–60]. However, such systems have limited effectiveness when objects are in the form of 3-D multi-layered clutter. Further, when objects are identical (like cylindrical pellets) without any distinctive features, problem becomes more complex. This requires the use of a three-dimensional (3-D) sensor [61–63]. Such sensors, for e.g.,

Table 2.1 Some approaches for the bin picking task and the proposed approach presented

Literature	Hardware used	Object	Approach/Algorithm
Matthias et al. [46]	Anno robotic arm and Kinect V2 camera	white coloured T-shaped object with visible silhouettes kept in Blue container	Point cloud segmentation using improved density-based spatial clustering of applications with noise (DBSCAN) algorithm, improved by combining the region growing algorithm as well as using Octree to speed up the calculation. The system then uses principle component analysis (PCA) for coarse registration and for fine registration, the iterative closest point (ICP) algorithm is used
Yan et al. [47]	Kinova 6-Dof robot and Ensenso industrial 3D camera	Large Corner-joint uniformly white in colour	6-DoF pose estimation system recognizes and localizes different types of workpieces. The depth image is the input and a set of 6-DoF poses for different types of objects is the output. The algorithm has an offline phase and an online phase. During the offline phase, the 3D model database is generated for recognition. In the online phase depth image is converted to point cloud, its processed and finally 6D pose is estimated and refined using ICP algorithm
Kleeberger et al. [48]	External 3D sensor mounted on each bin and an Ensenso N20-1202-16-BL stereo camera and a dual arm robot	Large Objects(gear, shafts etc.) with visible silhouettes and features	A new public dataset for 6D object pose estimation and instance segmentation for industrial bin-picking. The dataset comprises both synthetic and real-world scenes. For both, point clouds, depth images, and annotations comprising the 6D pose (position and orientation), a visibility score, and a segmentation mask for each object are provided. Along with the raw data, a method for precisely annotating real-world scenes is proposed
Proposed	KUKA KR5, Basler camera with Micro epsilon 2D laser scanner on end-effector	Small, uniformly black, textures and featureless cylindrical objects kept in a black metal container	1. Generation of 3D cloud from 2D scans from 2D sensor suitable for difficult to measure surfaces. 2. Low resolution 3-D scan to guide a localized search for suitable grasp surface to optimally utilize the information from multiple sensors. 2. Segmentation and pose estimation of featureless, symmetrical objects, uniform in colour without using learning based approach which tend to fail in such cases. 3. Fully functional online system for emptying of bin using a fusion of image and range data

SICK Ruler [64] are bulky, expensive and with lesser resolution. Hence an economic, less bulky and accurate sensor was sought. One can perform the same task using a camera and a 2-D sensor, namely, a laser scanner by appropriately calibrating them. Fusion of data from these two sensors, i.e., the camera and the laser scanner, not only allowed bin picking of the objects lying in multiple layers but also helped in quick detection of any disturbance caused by the robot.

The laser scanner used was Micro-epsilon ScanCONTROL 2900–100 with 12 μm resolution [65]. The scanner has the ability to provide 2-D information, namely, the depth and the distance of a pixel representing the measured object along the line of laser projection [66]. A schematic illustration of the setup in Fig. 2.2 is shown in Fig. 2.3. Note that $\mathbf{T}_L$ denotes the relationship between end-effector coordinate frame E and laser scanner coordinate frame $\{L\}$.

The approach is different from those proposed by others like [61–63], since a 2D laser scanner was used instead of a 3-D scanner. But it demands a calibration step to determine a relationship between the laser scanner and the robot's end-effector

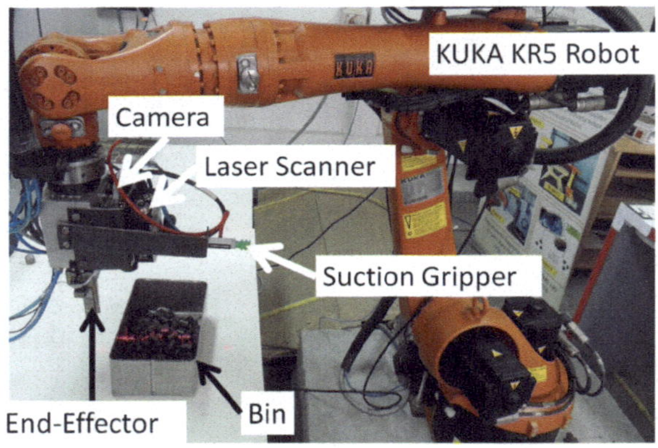

Fig. 2.2 A KUKA KR 5 arc robot with a camera and laser scanner mounted on its end-effector

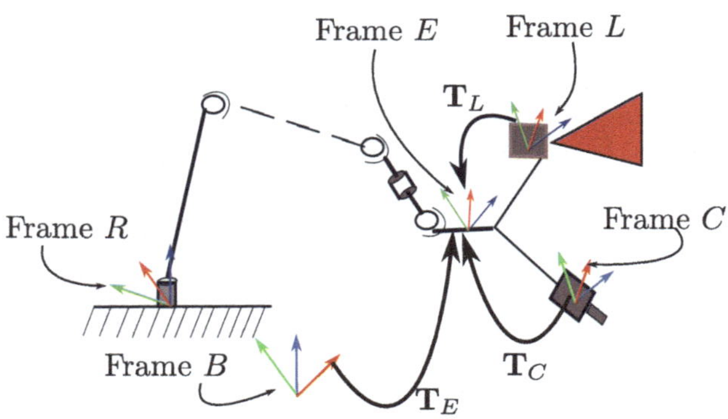

Fig. 2.3 Frames and transformations for identification

frames. Alternative method to do this is to correlate the scanner frame with the camera frame [67, 68], and then the camera frame to the end-effector frame [69]. Combination of these two gives the relationship of the scanner frame with respect to the end-effector frame. The latter method could not be used here due to requirement of higher accuracy [70], and will be further discussed in Sect. 2.2.4.

Conversion of the 2D data from the laser scanner to a single coordinate frame (B) is crucial for the applications that require 3D point cloud data. The calibration of the laser scanner to the camera was also important to enable the verification of the data emerging from the sensors.

Earlier approaches involved obtaining direct relationship of the transformation between the laser scanner's coordinate frame $\{L\}$ and the camera's coordinate frame $\{C\}$ [67, 68]. The method proposed by [68] required an initial 3D scan instead of 2D. This was not feasible using the 2D laser scanner discussed in this monograph.

It is pointed out here that the alternative approaches to calibrate frame $\{L\}$ to frame $\{C\}$ [67, 68] exist for mobile robots which do not require high accuracy. In these approaches, as well as the one proposed by [70], the information about the reference coordinate system is assumed to be correct. This is not so in the case of an industrial robot where the reference coordinate is defined on the end-effector whose pose itself has errors. In the bin picking application, however, higher accuracy (around few millimeters) was necessary. Hence, kinematic identification of the robot's geometric parameters is also necessary and will be discussed later in Chap. 4.

Therefore, a two-stage approach is proposed here for the calibration of the laser scanner with respect to the camera. In the first stage, kinematic calibration of the robot was undertaken to obtain all the actual kinematic (DH) parameters of the robot and the transformation between the end-effector's coordinate frame $\{E\}$ and the camera's coordinate frame $\{C\}$. It was followed by the calibration to find $\mathbf{T}_L$ the transformation between the end-effector's coordinate frame $\{E\}$ and the laser scanner's coordinate frame $\{L\}$. Both of them were then combined to obtain the transformation between camera and laser sensor coordinate frames. Note that calibration of robot end-effector w.r.t., its world frame referred as tool calibration is carried using four-point approach mentioned in robot's user manual and here in Appendix C. The identification of the transformation from end-effector's coordinate frame to laser scanner's coordinate frame is discussed later in this chapter.

2.2.2 *Camera*

The details of image formation while using a camera is presented in Appendix A.1. From Eqs. A.1 and A.2, it can be interpreted that use of a camera for measurement makes it necessary to identify the internal and external parameters of the camera. This process is called camera calibration [71]. Camera calibration is done using multiple correspondences. It may be done using a non-linear least square method. Optimization algorithms are used to obtain the solutions by refining an initial estimate

Fig. 2.4 Bin with cylindrical objects as pellets in cluttered fashion

[72, 73]. Multiple images of a calibration grid with known sizes of squares from specific viewing directions are required for identifying the optimum parameters [74].

After calibrating the camera intrinsic parameters, the next step is estimation of camera extrinsic parameters. This is commonly refered to as hand-eye calibration [69]. A 3D calibration grid with two planar calibration grids attached to one another was used used for accomplishing this. An external coordinate frame was defined on the calibration grid. The relative pose of the end-effector coordinate frame was used to represent the 3D grid coordinate points in the end-effector coordinate frame. Later this data was supplied along with the camera intrinsic parameter to the Efficient Perspective n-Point algorithm [75] implemented in OpenCV SDK. A bin with dark cylindrical objects is shown in Fig. 2.4. To automate such a task, an industrial robot mounted with suitable sensors can be used. This method resulted in good estimates using an image grabbed from a single pose of the camera mounted on the end-effector. More details are presented in [76].

2.2.3 *Laser Scanner*

Bin-picking automation in unstructured environments is challenging. One such task is commonly encountered in manufacturing industries for assembly etc. This is often done in a manual way.

Figure 2.2 shows a monocular camera and a 2D laser scanner mounted on the robot's end-effector for such an application. However, the setup requires the calibrations of the camera as well as the 2D laser scanner. In this chapter, calibration of the laser scanner is discussed such that the resulting setup can be used for bin-picking and some other applications.

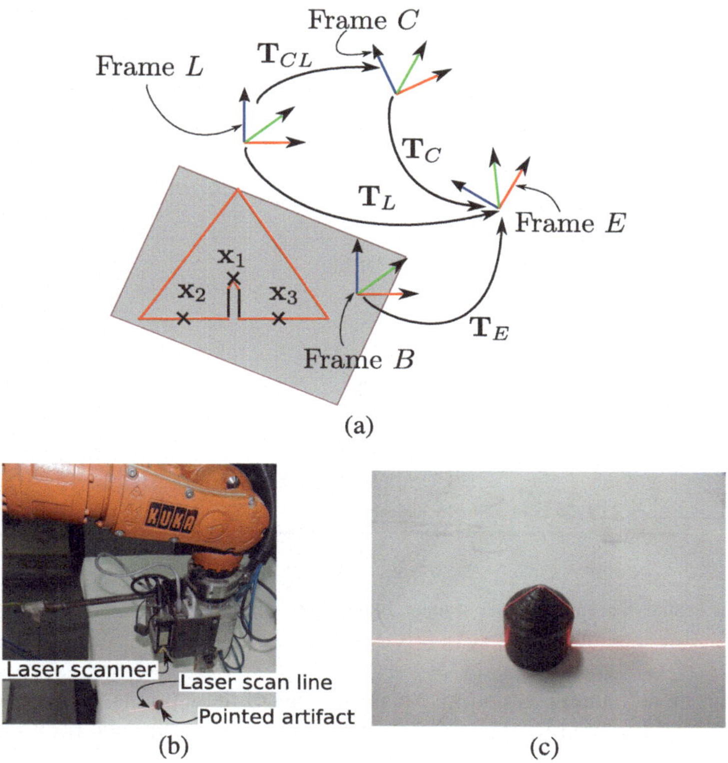

Fig. 2.5 Configuration of laser scanner frame ($\{L\}$) and robot end-effector frame ($\{E\}$), **a** Schematic, **b** Experimental setup, **c** Pointed artifact

2.2.4 Formulation

The schematic of the setup with three coordinate frames, namely, base frame $\{B\}$, end-effector frame $\{E\}$ and laser scanner frame $\{L\}L$ frames is shown in Fig. 2.5a. The transformation $\mathbf{T}_L$ between the frames E and L is to be calculated. The end-effector configuration $\mathbf{T}_E$ was calculated using the kinematic parameters of the robot. The base frame was defined using the method commonly used in industrial robots.

The identification was done in two stages. In the first stage, a rough estimate was made and this was refined in the final stage. A rough estimate ($\mathbf{T}_L^{est}$) of the transformation $\mathbf{T}_L$ was obtained in the following fashion. The laser scanner profile was projected on a flat surface and a pointed artifact[1] was manually positioned along the path of the profile, as shown in Fig. 2.5b. The profile obtained is shown in Figs. 2.5c and 2.6a. Three points on the profile are shown in Fig. 2.6b. Two lying on the plane

[1] Such an artifact can be easily made. The artifact shown in Fig. 2.5c was fabricated by machining a cylindrical piece made of steel in a lathe machine.

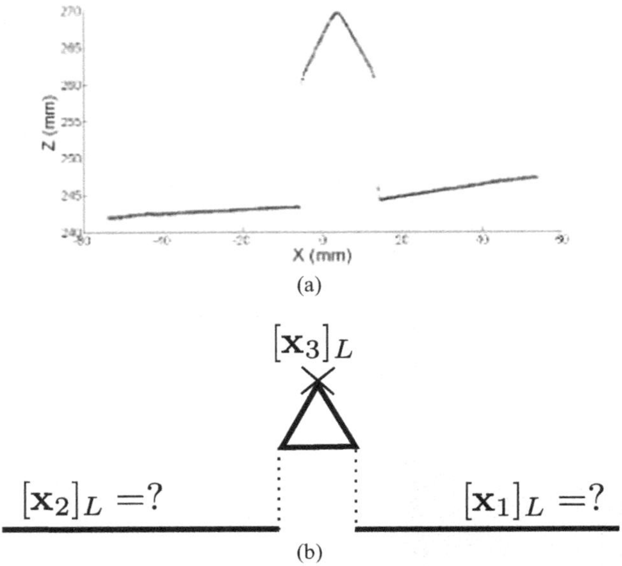

Fig. 2.6 **a** Profile of scan from laser scanner, **b** Schematic of 3-points

of measurement, namely, $\mathbf{x}_1$ and $\mathbf{x}_2$. Measurements of these points were made in the base coordinate frame (B). The third point was measured using the position of the pointed part of the artefact (Fig. 2.6a, b) in frame B as $[\mathbf{x}_3]_B$. The same point was calculated from the data obtained from the laser scanner frame L as $[\mathbf{x}_3]_L$. To obtain the rotation matrix $\mathbf{Q}_L^{est}$ of the transformation matrix $\mathbf{T}_L^{est}$ relating the frames E and L, the following steps were then followed.

Note that, the coordinates of $\mathbf{x}_1$ and $\mathbf{x}_2$ were available only in the base frame and are denoted by $[\mathbf{x}_1]_B$ and $[\mathbf{x}_2]_B$, respectively. All three points were transformed to the end-effector's frame (E) using $\mathbf{T}_E^{-1}$ as, $[\mathbf{x}_1]_E = \mathbf{T}_E^{-1}[\mathbf{x}_1]_B$, $[\mathbf{x}_2]_E = \mathbf{T}_E^{-1}[\mathbf{x}_2]_B$ and $[\mathbf{x}_3]_E = \mathbf{T}_E^{-1}[\mathbf{x}_3]_B$. All the three points in the profile are in the XZ plane of the laser scanner coordinate frame (L). By adjusting the robot's end-effector it was made sure that the points $[\mathbf{x}_1]_L$ and $[\mathbf{x}_2]_L$ have the same Z coordinates which mean that the line joining these points will approximately be parallel to the laser scanner's X axis. Therefore, the direction cosine of the X axis of the frame L is as follows:

$$\mathbf{r}_x = \frac{[\mathbf{x}_1]_E - [\mathbf{x}_2]_E}{||([\mathbf{x}_1]_E - [\mathbf{x}_2]_E)||} \tag{2.1}$$

Later, using the fact that $[\mathbf{x}_3]_L$ is in the XZ-plane, a unit vector on the XZ plane is given by

$$\mathbf{r}_{xz} = \frac{[\mathbf{x}_3]_E - [\mathbf{x}_1]_E}{||([\mathbf{x}_3]_E - [\mathbf{x}_1]_E)||} \tag{2.2}$$

Using Eqs. 2.1 and 2.2, the direction cosine of Y-axis is given by

$$\mathbf{r}_y = \frac{\mathbf{r}_{xz} \times \mathbf{r}_x}{||\mathbf{r}_{xz} \times \mathbf{r}_x||} \tag{2.3}$$

Finally, the direction cosine of Z-axis is as follows:

$$\mathbf{r}_z = \frac{\mathbf{r}_x \times \mathbf{r}_y}{||(\mathbf{r}_x \times \mathbf{r}_y)||} \tag{2.4}$$

The rotation matrix $\mathbf{Q}_L^{est}$ having the direction cosines as column was then obtained as:

$$\mathbf{Q}_L^{est} \equiv [\mathbf{r}_x \ \mathbf{r}_y \ \mathbf{r}_z] \tag{2.5}$$

To find the position of the origin of the coordinate frame L which is the translation part of $\mathbf{T}_L^{est}$, i.e, $\mathbf{t}_L^{est}$, measurements of $\mathbf{x}_3$ in both coordinate frames, i.e., L and B were used. Note that

$$[\mathbf{x}_3]_B = \mathbf{T}_E^{-1}\mathbf{T}_L^{est}[\mathbf{x}_3]_L \tag{2.6}$$

where $\mathbf{T}_E$ was known from the robot's controller, and $\mathbf{Q}_L^{est}$ was obtained in Eq. 2.5. Hence, one can find the vector $\mathbf{t}_L^{est}$, and accordingly obtain the transformation matrix $\mathbf{T}_L^{est}$. This is given by

$$\mathbf{T}_L^{est} \equiv \begin{bmatrix} \mathbf{Q}_L^{est} & \mathbf{t}_L^{est} \\ \mathbf{0}^{\mathsf{T}} & 1 \end{bmatrix} \tag{2.7}$$

To obtain an accurate value of $\mathbf{T}_L$, the pointed artifact was kept in multiple positions and measurements were taken in the frames B and L in the following fashion:

$$[\mathbf{x}]_B = \mathbf{T}_E^{-1}\mathbf{T}_L[\mathbf{x}]_L \tag{2.8}$$

These measurements were used to minimize the errors between the real and estimated values of the position.

To perform refinement of the parameters, the rotation matrix was converted to equivalent roll, pitch and yaw angles. Later, these angles were refined for minimum errors. Equation 2.8 was utilized to derive the expression used. The final estimate of the transformation matrix is as follows:

$$\mathbf{T}_L = \begin{bmatrix} -0.8232 & -0.5674 & -0.0202 & 114.024 \\ -0.5665 & 0.8233 & -0.0370 & 45.605 \\ 0.0377 & -0.0190 & -0.9991 & 340.280 \\ 0 & 0 & 0 & 1 \end{bmatrix}.$$

2.3 Segmentation and Pose Estimation

Image based segmentation performance is affected due to variation in lustre, lighting and lack of contrast, whereas 3-D segmentation is completely impervious to such factors. Thus, we begin our segmentation process with the point cloud generated from the depth map. Later, the results were processed by image based segmentation for missing information.

2.3.1 *3D Segmentation*

Once we have the input cloud from the scanner, we start processing the point cloud, by removing points corresponding outside points to container. The cloud is de-sampled to induce uniform sampling density and then filter statistical outliers as described in [77]. A uniform spatial density, ensures better determination of geometric properties. Only points corresponding to the pellets remain in the cloud after this process.

2.3.2 *Mathematical Formulation for Pose Estimation*

Let a point be represented by $\mathbf{p} = [x\ y\ z]^T \in \mathbb{R}^3$. Let there be m such points in a cluster. Then, the cluster can be represented as $\mathrm{P} = \bigcup_{i=1}^{m} \mathbf{p}_i(x_i, y_i, z_i)$. The normals (Fig. 2.8b) at each point is represented by $\mathbf{n}_i$ and is computed by fitting a plane to points in its neighbourhood. The neighbourhood criteria is detailed in [77]. The Principle Curvature Information (PCI), for each point, is computed using [77] and is represented $\mathbf{c}_i$. This is found by applying Principal Component Analysis (PCA) on the components of the direction of the estimated normal for each point.

From each point an estimate for the axis is computed, refer Fig. 2.8c, given by $\mathbf{a}_i = \mathbf{n}_i \times \mathbf{c}_i$. This gives us more than 100 samples of the same variable. We chose our estimate for axis value chosen as the X-axis of the object frame as:

$$\mathbf{a}^{est} = \frac{\sum_{i=1}^{m}(\mathbf{n}_i \times \mathbf{c}_i)}{m} \tag{2.9}$$

Consider that $\forall\ \mathbf{p}_i \in \mathrm{P}$, the coordinates in Base Frame are given by $[\mathbf{p}_i]_B = [x_i\ y_i\ z_i]^T$ and co-ordinates of $\mathbf{p}_i$ in Localized Object Frame of the cluster are given by $[\mathbf{p}_i]_O = [x_i{}^O\ y_i{}^O\ z_i{}^O]^T$. Then, using the scalar product operator ($<.>$), x_i^0 is obtained as:

$$x_i{}^0 = <\mathbf{p}_i, \mathbf{a}^{est}> \tag{2.10}$$

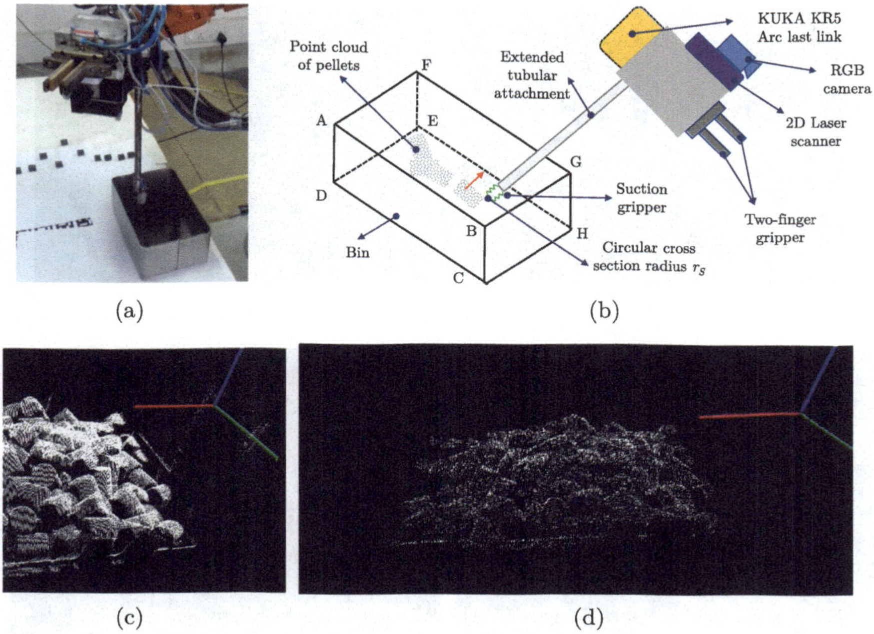

Fig. 2.7 **a** End-effector mountings **b** line diagram of the arrangement of sensors and bin **c** Data acquired using Laser scanner, **d** Filtered data

Additionally, top of the cylinder as $t_c = max(x_i{}^0)$ and base of the cylinder $b_c = min(x_i{}^0)$. Also, centroid of the cylinder along x-axis in the object frame is calculated as $m_c = \overline{x_i{}^0}$. Height of the cylinder $h_c = t_c - b_c$ (Fig. 2.7).

Now, if the minimum un-occluded height of the cylinder is greater than radius of the suction gripper, r_S. Then, let $\mathbb{C} = \{\mathbf{p}_i \in \mathrm{P} \parallel \; |x_i{}^0 - m_c| < r_S\}$. Since we would initiate pickup only if, cylinder is un-occluded around the pick-up point, thus we project only points in $\mathbb{C}$ onto a plane passing through the center of the cylinder and perpendicular to axis, using a map $\mathtt{P} : \mathbb{C} \to \mathbb{R}^3$

$$\mathtt{P}(\mathbf{p}_i) = \mathbf{p}_i - ((< \mathbf{p}_i, \mathbf{a}^{est} > -m_c) \times \mathbf{a}^{est}) \tag{2.11}$$

$$\mathtt{P}(\mathbf{p}_i) \qquad = \mathbf{p}_i - ((x_i{}^0 - m_c)\mathbf{a}^{est}) \tag{2.12}$$

Applying the transformation as $[\mathbf{P}]_B^O[x_i, y_i, z_i]_B$ will result in $[m_c, y_i{}^O, z_i{}^O]_O^T$, which gives a circular arc in the central plane perpendicular to the axis of the cylinder, as shown in Fig. 2.8d.

Applying PCA on the estimated normals, we get the centroid of the cylinder as vector $[\mathbf{g}]_B^{est}$ represented as (x_c, y_c, z_c), and the three Eigen vectors as $[\mathbf{e}_1, \; \mathbf{e}_2, \; \mathbf{e}_3]$ in order of decreasing significance and the thre variance which are Eigne values corresponding to the three Eigen vector as $[\upsilon_1, \; \upsilon_2, \upsilon_3]$ respectively.

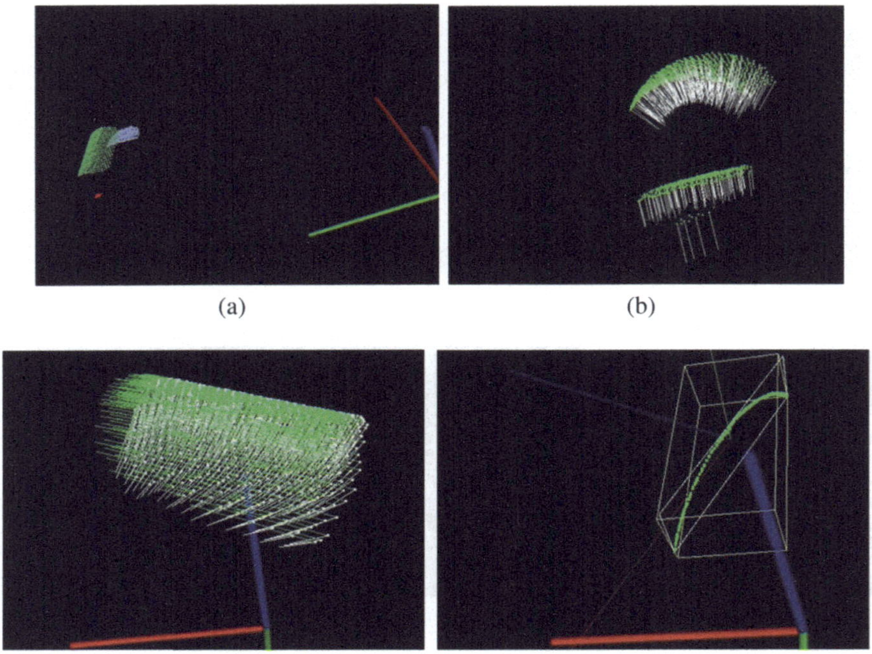

(a) (b) (c) (d)

Fig. 2.8 **a** Region growing in voxel, **b** Normals estimation, **c** Axis calculated, **d** Cylindrical cluster projected onto the plane axis

For the object frame, axis $\mathbf{x}_B^{est} \equiv \mathbf{e}_3$, is along the cylinder axis. Axis $\mathbf{y}_B^{est} \equiv \mathbf{e}_1$, is along the tangent and axis $\mathbf{z}_B^{est} \equiv \mathbf{e}_2$, is along the approach direction for pickup.

Also, if $v_1 > 2r_S$ and $v_3 > 2r_S$ then the object is *approachable*. The size of the object as estimated in the object frame was then compared with the actual dimension of the object. Three conditions may arise:

- If size is within acceptable deviation, then the object is qualified and the estimates are supplied for pickup.
- Due to heavy occlusion the size of the patch is smaller than expected. Based on how small it is the patch was either discarded or forward for image based verification.
- In cases when multiple pellets were nearly aligned with one another, the length and breadth exceeded the expected value. These cases were also carried forward to image based refinement.

Algorithm 1 3-D Segmentation

Input: Conditioned Point cloud data from Laser scanner .

Output: Surface Centroid and Approach Normal. *LOOP Process* :

1: **Normal estimation:** Normals were estimated corresponding to each point in the Voxel, as shown in Fig. 2.8a, b

2: **Curvature estimation:** Variation in direction of normals of neighbors, was associated with curvature of each point.

3: **Region growing:** Patches were found using a region growing algorithm [77] which utilizes curvature value, direction of normals and euclidean distance between each point.

4: **Classification:** Each patch was classified to be belonging to specific shape (in our case plane or cylinder), based on mean curvature value.

5: Centroid: Mean of position of all points was the estimate for patch centroid.

6: **If** Surface is flat : Approach normal: The average of direction of normals gave us the approach normal to the surface.

7: **Elseif** : Surface is cylindrical

8: **Tangents and normals**: For each point the direction of maximum deviation of normals gave us the direction of the tangent as shown in Fig. 2.8c. Using the cross product of tangent and normal at each point the mean estimate for axis was computed.

9: **Cross-sectional plane**: All points on the cylinder patch were projected on the plane passing through the axial center and with the axis as normal, Fig. 2.8d.

10: **Axis and approach normal**: Applying PCA to these points we got axis as the eigenvector with smallest eigenvalue and approach normal as the eigenvector with the second highest eigenvalue.

11: **Object bounding box**: The length and breadth of the patch were also calculated by finding variation in projected lengths along basis vector of object frame.

return Surface patch, centroid, approach normal, axis.

2.3.3 Image segmentation

Thus image based segmentation is required to tackle cases of severe occlusion and when surface of closely associated objects is in alignment as shown in Fig. 2.9a, b. The point cloud processing provides us with localized surface patch with known surface

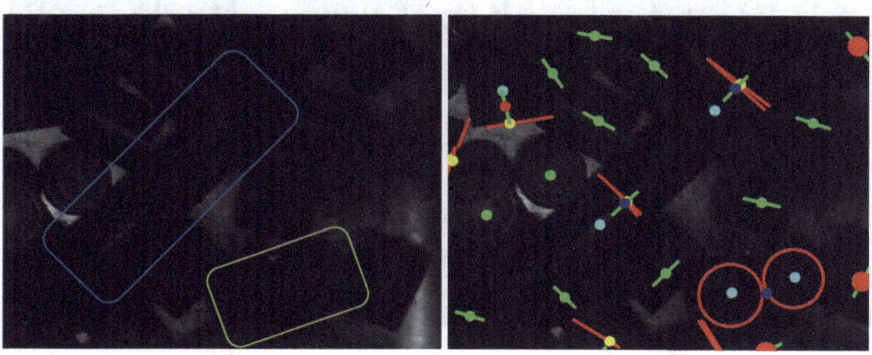

(a) (b)

Fig. 2.9 **a** Image for aligned cylinders, **b** Segmentation results type and orientation of this patch. Limitations observed in image based segmentation are resolved as follows :

- Low relative contrast is improved by applying histogram equalization on the Region of Interest (ROI) which is decided based on 3D segmentation .
- Detection of false edges due to specularity is circumvented by prior knowledge of expected features. Using the information generated from 3D segmentation we search for specific information within the image, this helps reduce false detections.

Algorithm 2 Image Segmentation

Input: Surface patch, type of surface, centroid, approach normal, axis, bounding box in object frame.
Output: Center.
1: *LOOP Process* :
2: **Projection:** Points from the surface patch are projected onto the image.
3: **Mask :** Using this projection an image mask is created to define a ROI.
4: **Histogram equalisation:** In the selected ROI histogram equalisation is applied to locally enhance contrast.
5: Based on surface classification from 3D :
6: **If** Surface is flat
7: **Circles:** Circles are fit to the equalized image and center is estimated. This provides the top of cylinder in the voxel, for results see Fig. 2.10a, b.
8: **Else if** Surface is cylindrical
9: **Lines:** Then the axis calculated from 3D data is projected onto the image. Lines are fit in a direction perpendicular to that of the axis and center is triangulated using half the projected length of the axis.
10: **return** Projected center

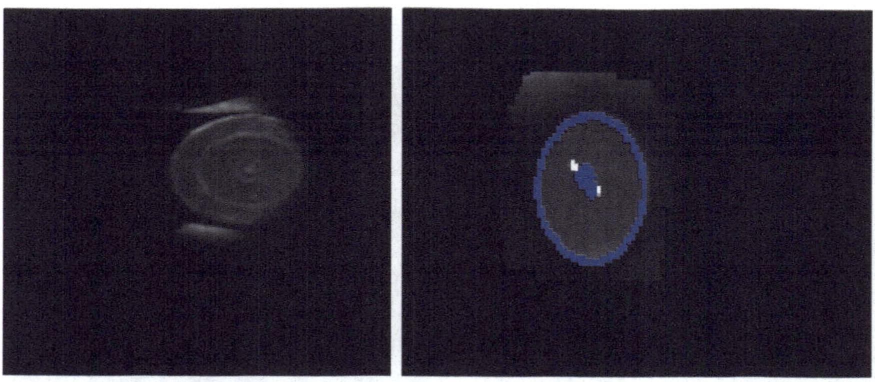

(a) (b)

Fig. 2.10 After extracting cluster from 3D **a** shows region of interest and **b** show segmentation result in standing pellet case

As opposed to when applied on complete image histogram equalization of the localized patch, greatly increases the contrast. This provides the requisite detail for successful image segmentation and also imbues immunity to variance in lighting conditions. By looking for specific features in localized region of high contrast , consistent results are achieved. This center obtained from the image is back projected onto the cloud. Using the estimates of orientation and normal for approach, acquired from 3D data, pellet pose and position are calculated (Fig. 2.10).

The key insight is to enhance contrast by localized processing and using the surface type and object orientation to guide the image search. Proposed algorithm has shown reliable performance and is resilient to change in ambient lighting.

2.4 Practical Considerations for Pick-Up Using Robotic Setup

On conditioning the data, received from the sensor, the cylindrical pellets were found and orientation was estimated by the Algorithm 1 and its mathematical formulation in Sect. 2.3.2. The image segmentation algorithm is listed in Algorithm 2. Pellets were uniformly black in color and cylindrical in shape. They are 14–17 mm high with a radius of 8 mm upto 0.5 mm. They were piled in a stainless steel container, as much as 8 cm high. By varying these parameters, the implementation could be extended to cylindrical objects of different sizes. Graspability was ascertained by comparing patch dimensions in object frame with the gripper diameter. If qualified the surface centroid and approach normal information were transferred to KUKA robot in the form of position and orientation. The Z-axis of the gripper rod was aligned with the estimated approach normal and X-axis was aligned with the estimated axis of the object while picking it up. Moreover the collision detection avoidance using the geometric approach is detailed in Chap. 6.

2.4.1 Disturbance Detection

After each pick-up, an image is taken of the container from the same perspective. This image was compared with the previous image. Pixel by pixel difference was calculated in both images, as shown in Fig. 2.11b. If the pixel difference was found to be above the noise threshold decided by repeated trials as shown in Fig. 2.11a the change was estimated to be caused due to movement or disturbance in the scene. Range data corresponding to this pixel was deleted using the method of backward projection.This ensured that these points were not used in the segmentation process of the next pellet. Thus, the disturbed pellets were successfully ignored while circumventing the need of a re-scan.

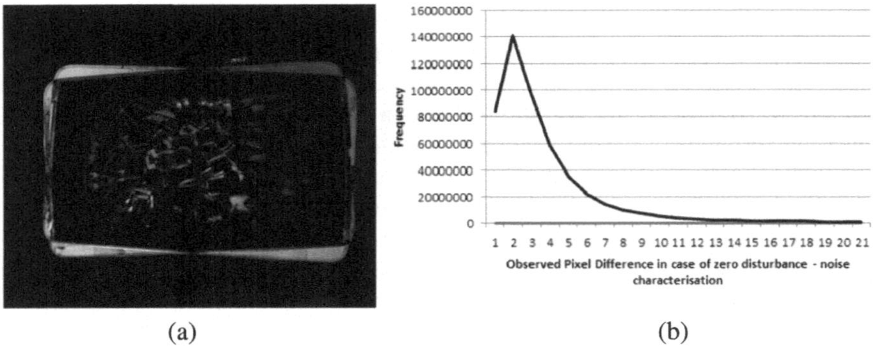

Fig. 2.11 **a** Pixel difference observed without disturbance, **b** Visualization of disturbance

During pick-up, there were other sources of disturbance besides collision with the container which were hard to model. These may be caused due to object-object interaction, or may be due to the object and gripper interaction. To cater to such disturbances without initiating a fresh scan, another image was taken after pick-up of the object. This allows multiple pick-ups based on information of a single scan. The parameters specific to the method and criterion were discussed earlier.

2.4.2 *Oriented Placing*

The picked objects were placed in uniform pose using the estimate of orientation and keeping in consideration the approximations made to make the approach feasible.

2.4.3 *Shake Criterion*

We considered that a cycle begins from a scan and terminates at the requirement of another scan. Multiple pellets were picked up in a cycle. Some pellets were not picked up to avoid collision between gripper and the container. such an attempt may result in misalignment of all pellets. If the number of pellets picked up in a given cycle, was very less as compared to the total number of pellets segmented in the cycle, then the boat shake is done. his is owing to the reason that given the situation most of the objects may not be grasped due to the possibility of gripper to bin collision. The condition is imposed as the time cost per pellet for conducting a scan increase as the number of pellets picked in a cycle decreases.

2.5 Experiments and Results

The results of the laser scanner to end-effector calibration is discussed in this section.

2.5.1 Laser Scanner to End-Effector

Accuracy of calibration for the laser scanner was determined in the following fashion. The pointed artifact, which was used earlier for calibration, was kept on a calibration grid print out whose grid sizes were known (Fig. 2.12). The diameter of the object as well as its height was measured using a calliper of resolution 0.02 mm. The object was placed in such a way that knowing the grid size, diameter, and height of the object, the position of the tip will be known in the grid coordinate frame i.e., base coordinate frame $\{B\}$. The relationship between this frame and the coordinate frame $\{R\}$ was measured using the methodology implemented in robot controllers. Later, during the measurement stage, the robot was positioned such that the laser scan line passes through the exact top-most point of the object. Following this, measurement of the top-most point was made in laser scanner coordinate frame ($\{L\}$). It was then converted to the calibration grid coordinate frame using the calibrated parameters, as obtained in Sect. 2.2.4. The norm of the differences between that measured using the laser scanner and the true values was calculated. The experiments were conducted multiple times (here, 25 times) and the mean is reported in Table 2.2.

The corresponding values obtained using the proposed method was compared with those from the methods by [67, 70]. They are also shown in Table 2.2. While using the method proposed by [70], the nominal parameters of the robot was used.

Fig. 2.12 Accuracy measurement (d_a and h_a are diameter and height of the artifact)

Table 2.2 Performance of calibration

Method	$\{L\}$ to $\{C\}$ [67]	$\{L\}$ to $\{E\}^{\dagger}$	
		Roy et al. [70]	Proposed
Accuracy (mm)	16	2.548	1.021

$\{L\}$: laser scanner coordinate frame; $\{C\}$: camera coordinate frame; $\{E\}$: end-effector coordinate frame. †Average results over 25 points

It may be noted that the results of the method proposed by [67] was obtained using codes that are available online [78]. It is clearly evident that the performance of the proposed method is better than the rest of the methods. Additional observation is that the improvement in the computed value while using calibrated and uncalibrated robot is about 1.5 mm. Therefore, it can be inferred that the camera based identification of the robot had a profound effect on the calibration of the laser scanner also.

2.5.2 *Transformation Between Laser Scanner and Camera*

Figure 2.13 depicts the performance of the proposed calibration methodology. The figures were created in the following way. The calibration grid was first scanned using the laser scanner mounted on the robot to obtain the 3D point clouds of the planar surface of the grid. The calibrated parameters of the robot as well as the laser scanner to end-effector transformation were utilized to do so, such that the 3D scan was obtained. An image of the calibration grid was also captured. Once the 3D points were obtained in the base frame B, these points were projected onto the image using the calibrated values of the robot and the camera to end-effector transformation. In short, Fig. 2.13a is a representation of the performance of all the parameter identification discussed in this chapter, namely, camera calibration, camera to end-effector calibration, and finally the laser scanner to end-effector calibration. By adjusting the exposure settings, the laser scanner was made to detect only the planar points on the white squares of the grid. In Fig. 2.13a, it may be noted that the scan points are correctly projected only on the white squares from where the 3D points were obtained. The result of the transformation using the method proposed here is qualitatively illustrated with the projection of 3D points from the laser scanner on the image itself as in Fig. 2.14.

Once the optimum values of $\mathbf{T}_C$ and $\mathbf{T}_L$ were obtained, the relationship between frames C and L, namely, $\mathbf{T}_{CL}$, are easily obtained as

$$\mathbf{T}_{CL} = \mathbf{T}_C\mathbf{T}_L^{-1}. \tag{2.13}$$

To have a clearer idea of the accuracy of this transformation, a single line scan of a pointed artifact was obtained. At the same pose of the robot, an image of the scan line was also grabbed. The obtained 3D coordinates was then projected on this image, as shown in Fig. 2.13b. It is worth noting that the (blue) projected line is exactly matching with the laser scan line. At the same time, the parameters obtained using the method proposed by [70] showed deviation, as visible in Fig. 2.13c.

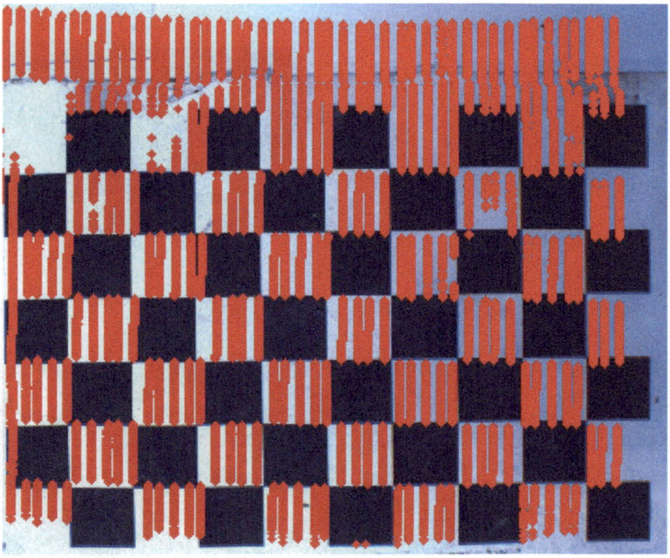

(a)

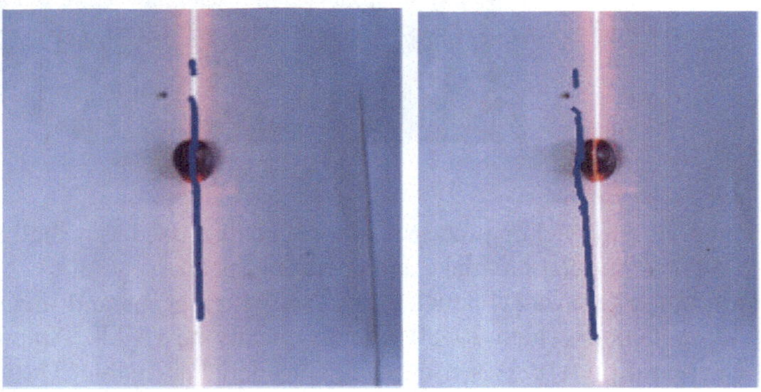

(b) (c)

Fig. 2.13 **a** Projection of 3D scan points on calibration grid image, **b** Laser line scan on its image using parameters from proposed method **c** line scan on its image using parameters from [70]

2.5.3 Segmentation and Pose Estimation

Pellets were placed in the bin in multi-layer configuration and random orientations. Multi layer configuration was attempted to test cases of severe occlusion by piling up pellets in heap. Average scan time was 8 seconds. Though scan timing would vary based on equipment and robot configuration, it is mentioned to put in perspective the time that is saved by reducing the number of scans required for emptying the

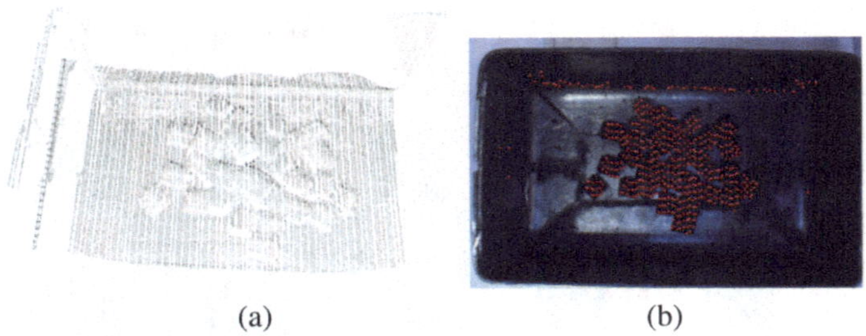

Fig. 2.14 **a** Illustration showing 3D scan points of bin, **b** projection of 3D points on image

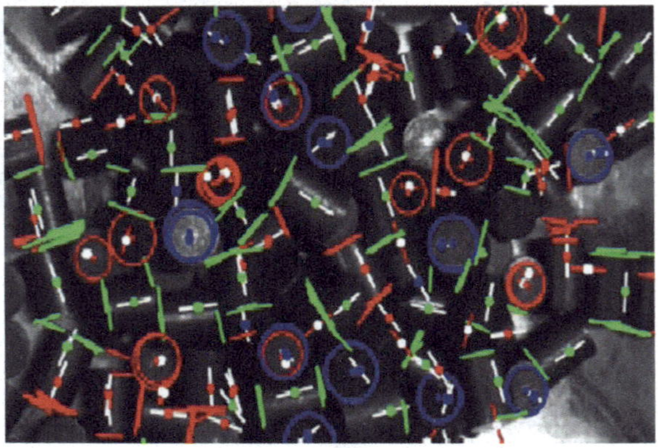

Fig. 2.15 Green colour represents of surfaces as classified by 3D analysis. Red colour is associated when results from 3D segmentation are non-conclusive

bin. Complete timing for pickup process has not been discussed as it shall be highly subjected to robot configuration, make and operating speeds.

The segmentation test was conducted on a set of ten scans and images. Each containing 125 pellets of which 70–80 would typically be visible. Among all the objects 40% exposed (in a fuzzy sense), were considered as visible. The algorithm successfully determined the position and orientation of 70% of these visible objects. While dedicatedly running on single threaded the process took 300 ms to isolate and find the attributes for a single pellet. High detection rate in a single scan and real time pose determination allow locating multiple pellets utilizing the time required for robot motion as shown in Fig. 2.15. It may be possible that if some objects are found to be in blind spot for the manipulator which allows to properly plan the process. This eliminates processing delays and also reduces on requirement of multiple scans for consecutive pickup.

Domae et al. [63] studied objects of similar scale, reporting a segmentation time of is 270 ms. Though the algorithms does not determine the pose or orientation in stated time and pick-up is done in restriction of 4 DOF. Buchholz et al. [62] utilizes RANSAC which is computationally intensive and reports a timing of 4000 ms. Though notably objects are bigger in size and of higher complexity, iterative schemes are known to have high computational requirements. Matthias et al. [79] uses optimized RANSAC notes a timing of 2000 ms for primitive objects, though of bigger size. They also report pose estimation of roughly 50% of the objects visible (fuzzy measure). Liu et al. [80] succeeds in determining the pose in 900 ms and with high detection rate. The objects were of size similar to what is discussed here but the algorithm would suffer from limitations stated in the introduction for purely depth based algorithms.

2.5.4 System Repeatability

When running the whole algorithm for emptying of the bin. Gripper approach for pickup is coincident with estimated normal to the surface at the calculated centroid. Success rates was 93% in 1200 pick-ups with the failures cases include misclassification between planar and curved surface. In case of failure, 2%—pickup is successful but orientation is missed. In addition to this, 2% are associated to segmentation failure, for example, centroid estimate of the exposed portion is too close to the boundary of the object. Finally, 3% of misses are attributed to disturbance of object while pick-up process due to gripper movement and suction failure. Please note the objects are in a pile and therefore are highly unstable. In such as situation even slight disturbance causes shift in object position. Domae et al. [63] specifically focuses on determining grasp-ability. Though it does so in 4-DOF. They have reported accuracy of 92% though it appears they have most objects in simple orientation, parallel to the base plane. During experimentation, the pick-ups were prone to failure in case an angled approach is not implemented for tilted objects, more so for curved/ non planar surfaces. A pickup accuracy of 99% is reported in [62]. the objects illustrated seem to be in relatively low clutter and have planar surfaces. Which significantly improves performance due to lack entanglement, improved object stability and generally more approachability due to low occlusion. Matthias et al. [79] 90% from 32 attempts accuracy reported, reasons cited for failure are similar as most observed—entanglement, collision with neighbouring object while pick-up. Liu et al. [80] reported 94% pick-up accuracy. The failure cases were attributed to occlusion and gripper failure. Table 2.3 lists the timing and efficiency of the proposed method with the one discussed in literature as cited.

Table 2.3 Timing and efficiency as compared to state of the art (SOA) algorithms

SOA approach	Timing ms	Efficiency (%)
Domae et al. [63]	270	92
Buchholz et al. [62]	4000	99
Matthias et al. [79]	2000	90
Liu et al. [80]	900	94
Proposed	300	93

2.6 Summary

This chapter discussed about the calibration of a camera and 2D laser sensor which are mounted on an industrial robot for bin picking application. Deterministic approach to multi-sensor based segmentation techniques is also showcased in this chapter. The segmentation and pose determination algorithms were built based on information generated by the 3D segmentation process to refine the search space for the image segmentation. The algorithm though depending upon RGB data shows immunity to illumination variance by virtue of localised image processing. The segmentation counters the limitations posed by the effects like specularity, axial alignment, facial alignment, and low contrast. We hope that this initiates a new class of algorithms which would look for specific information from different sensory inputs. Utilising the strength of the individual sensor to improve overall performance of the system.

References

1. Zisserman, A., Forsyth, D., Mundy, J., Rothwell, C., Liu, J., and Pillow, N.: 3d object recognition using invariance. Artif. Intell. **78**(1), 239–288 (1995). Special Volume on Computer Vision
2. Ponce, J., Chelberg, D., Mann, W.B.: Invariant properties of straight homogeneous generalized cylinders and their contours. IEEE Trans. Pattern Anal. Mach. Intell. **11**(9), 951–966 (1989)
3. Zerroug, M., Nevatia, R.: Three-dimensional descriptions based on the analysis of the invariant and quasi-invariant properties of some curved-axis generalized cylinders. IEEE Trans. Pattern Anal. Mach. Intell. **18**(3), 237–253 (1996)
4. Faugeras, O., Hebert, M.: The representation, recognition, and locating of 3-d objects. **5**, 27–52 (1986)
5. Ansar, A., Daniilidis, K.: Linear pose estimation from points or lines. IEEE Trans. Pattern Anal. Mach. Intell. **25**(5), 578–589 (2003)
6. Johnson, A.E., Hebert, M.: Using spin images for efficient object recognition in cluttered 3d scenes. IEEE Trans. Pattern Anal. Mach. Intell. **21**(5), 433–449 (1999)
7. Mittrapiyanumic, P., DeSouza, G.N., Kak, A.C.: Calculating the 3d-pose of rigid-objects using active appearance models. In: 2004 IEEE International Conference on Robotics and Automation, 2004. Proceedings. ICRA '04, vol. 5, pp. 5147–5152 (2004)
8. Ling, H., Jacobs, D.W.: Shape classification using the inner-distance. IEEE Trans. Pattern Anal. Mach. Intell. **29**(2), 286–299 (2007)
9. Berg, A.C., Berg, T.L., Malik, J.: Shape matching and object recognition using low distortion correspondences. In: 2005 IEEE Computer Society Conference on Computer Vision and Pattern Recognition (CVPR'05), vol. 1, pp. 26–33 (2005)
10. Ferrari, V., Tuytelaars, T., Van Gool, L.: Object detection by contour segment networks. In: Proceedings of the 9th European Conference on Computer Vision—Volume Part III, ECCV'06, pp. 14–28. Springer, Berlin
11. Ferrari, V., Fevrier, L., Jurie, F., Schmid, C.: Groups of adjacent contour segments for object detection. IEEE Trans. Pattern Anal. Mach. Intell. **30**(1), 36–51 (2008)
12. Ferrari, V., Jurie, F., Schmid, C.: From images to shape models for object detection. Int. J. Comput. Vision **87**(3), 284–303 (2010)
13. Qi, C.R., Liu, W., Wu, C., Su, H., Guibas, L.J.: Frustum pointnets for 3d object detection from rgb-d data. In: Proceedings of the IEEE Conference on Computer Vision and Pattern Recognition, pp. 918–927 (2018)

14. Qi, C.R., Su, H., Mo, K., Guibas, L.J.: Pointnet: Deep learning on point sets for 3d classification and segmentation. In: Proceedings of the IEEE Conference on Computer Vision and Pattern Recognition, pp. 652–660 (2017a)
15. Lowe, D.G.: Distinctive image features from scale-invariant keypoints. Int. J. Comput. Vision **60**(2), 91–110 (2004)
16. Suga, A., Fukuda, K., Takiguchi, T., Ariki, Y.: Object recognition and segmentation using sift and graph cuts. In: 2008 19th International Conference on Pattern Recognition, pp. 1–4. IEEE (2008)
17. He, K., Gkioxari, G., Dollár, P., Girshick, R.: Mask r-cnn. In: Proceedings of the IEEE International Conference on Computer Vision (2017)
18. Liu, W., Anguelov, D., Erhan, D., Szegedy, C., Reed, S., Fu, C.-Y., Berg, A.C.: Ssd: Single shot multibox detector. In: European Conference on Computer Vision, pp. 21–37. Springer, Berlin (2016)
19. Redmon, J., Farhadi, A.: Yolo9000: better, faster, stronger. In: Proceedings of the IEEE Conference on Computer Vision and Pattern Recognition, pp. 7263–7271 (2017)
20. Badrinarayanan, V., Kendall, A., Cipolla, R.: Segnet: A deep convolutional encoder-decoder architecture for image segmentation. IEEE Trans. Pattern Anal. Mach. Intell. **39**(12), 2481–2495 (2017)
21. Qi, C.R., Su, H., Mo, K., Guibas, L.J.: Pointnet: Deep learning on point sets for 3d classification and segmentation. In: Proceedings of the IEEE Conference on Computer Vision and Pattern Recognition, pp. 652–660 (2017b)
22. Li, Y., Bu, R., Sun, M., Wu, W., Di, X., Chen, B.: Pointcnn: Convolution on x transformed points (2018). ArXiv:180107791
23. Dong, Z., Liu, S., Zhou, T., Cheng, H., Zeng, L., Yu, X., Liu, H.: Ppr-net: point-wise pose regression network for instance segmentation and 6d pose estimation in bin-picking scenarios. In: 2019 IEEE/RSJ International Conference on Intelligent Robots and Systems (IROS), pp. 1773–1780. IEEE (2019)
24. Drost, B., Ulrich, M., Navab, N., and Ilic, S.: Model globally, match locally: Efficient and robust 3d object recognition. In: 2010 IEEE Computer Society Conference on Computer Vision and Pattern Recognition, pp. 998–1005. IEEE (2010)
25. Kruglyak, L., Lander, E.S.: Complete multipoint sib-pair analysis of qualitative and quantitative traits. Am. J. Human Gene. **57**(2), 439 (1995)
26. Vidal, J., Lin, C.-Y., and Martí, R.: 6d pose estimation using an improved method based on point pair features. In: 2018 4th International Conference on Control, Automation and Robotics (iccar), pp. 405–409. IEEE (2018)
27. Xiang, Y., Schmidt, T., Narayanan, V., Fox, D.: Posecnn: A convolutional neural network for 6d object pose estimation in cluttered scenes (2017a). ArXiv:1711.00199
28. Sock, J., Kim, K.I., Sahin, C., Kim, T.-K.: Multi-task deep networks for depth-based 6d object pose and joint registration in crowd scenarios (2018). arXiv:1806.03891
29. Zeng, A., Yu, K.-T., Song, S., Suo, D., Walker, E., Rodriguez, A., Xiao, J.: Multi-view self-supervised deep learning for 6d pose estimation in the amazon picking challenge. In: 2017 IEEE International Conference on Robotics and Automation (ICRA), pp. 1386–1383. IEEE (2017)
30. Bicchi, A., Kumar, V.: Robotic grasping and contact: A review. In: Proceedings 2000 ICRA. Millennium Conference. IEEE International Conference on Robotics and Automation. Symposia Proceedings (Cat. No. 00CH37065), vol. 1, pp. 348–353. IEEE (2020)
31. Caldera, S., Rassau, A., Chai, D.: Review of deep learning methods in robotic grasp detection. Multimodal Technol. Interact. **2**(3), 57 (2018)
32. Kuo, H.Y., Su, H.R., Lai, S.H., Wu, C.C.: 3d object detection and pose estimation from depth image for robotic bin picking. In: 2014 IEEE International Conference on Automation Science and Engineering (CASE), pp. 1264–1269 (2014)
33. Vosselman, G., Dijkman, S.: 3D building model reconstruction from point clouds and ground plans, pp. 37–43. Number XXXIV-3/W4 in (International Archives of Photogrammetry and Remote Sensing : IAPRS : ISPRS. International Society for Photogrammetry and Remote Sensing (ISPRS) (2001)

34. Papazov, C., Haddadin, S., Parusel, S., Krieger, K., Burschka, D.: Rigid 3d geometry matching for grasping of known objects in cluttered scenes. I. J. Robot. Res. **31**(4), 538–553 (2012)
35. Nieuwenhuisen, M., Droeschel, D., Holz, D., Stückler, J., Berner, A., Li, J., Klein, R., Behnke, S.: Mobile bin picking with an anthropomorphic service robot. In: 2013 IEEE International Conference on Robotics and Automation, pp. 2327–2334 (2013)
36. Rabbani, T., Heuvel, F.V.D.: Efficient hough transform for automatic detection of cylinders in point clouds (2005)
37. Bolles, R.C., Horaud, P.: 3DPO: A Three-Dimensional Part Orientation System, pp. 399–450. Springer US, Boston, MA (1987)
38. Lu, F., Milios, E.E.: Robot pose estimation in unknown environments by matching 2d range scans. In: 1994 Proceedings of IEEE Conference on Computer Vision and Pattern Recognition, pp. 935–938 (1994)
39. Latecki, L.J., Lakaemper, R., Sun, X., Wolter, D.: Building polygonal maps from laser range data (2004)
40. Baillard, C., Schmid, C., Zisserman, A., Fitzgibbon, A., England, O.O.: Automatic line matching and 3d reconstruction of buildings from multiple views. In: ISPRS Conference on Automatic Extraction of GIS Objects from Digital Imagery, IAPRS vol. 32, Part 3-2W5, pp. 69–80 (1999)
41. Curless, B., Levoy, M.: A volumetric method for building complex models from range images. In: Proceedings of the 23rd Annual Conference on Computer Graphics and Interactive Techniques, SIGGRAPH '96, pp. 303–312, ACM, New York, NY, USA (1996)
42. Weingarten, J.W., Gruener, G., Siegwart, R.: A state-of-the-art 3d sensor for robot navigation. In: 2004 IEEE/RSJ International Conference on Intelligent Robots and Systems (IROS) (IEEE Cat. No.04CH37566), vol. 3, pp. 2155–2160 (2004)
43. Gummadi, V.M. Sarkodie-Gyan, T.: 2 d object recognition using the hough transform. In: Intelligent Systems Design and Applications, pp. 413–421. Springer, Berlin (2003)
44. Saxena, A., Driemeyer, J., Ng, A.Y.: Learning 3-d object orientation from images. In: 2009 IEEE International Conference on Robotics and Automation, pp. 794–800. IEEE (2009)
45. Sundermeyer, M., Marton, Z.-C., Durner, M., Triebel, R.: Augmented autoencoders: Implicit 3d orientation learning for 6d object detection. Int. J. Comput. Vision **128**(3), 714–729 (2020)
46. Matthias, N., Droeschel, D., Holz, D., Stückler, J., Berner, A., Li, J., Klein, R., Behnke, S.: Mobile bin picking with an anthropomorphic service robot. In: 2013 IEEE International Conference on Robotics and Automation, pp. 2327–2334 (2013c)
47. Yan, W., Xu, Z., Zhou, X., Su, Q., Li, S., Wu, H.: Fast object pose estimation using adaptive threshold for bin-picking. IEEE Access **8**, 63055–63064 (2020)
48. Kleeberger, K., Landgraf, C., Huber, M.F.: Large-scale 6d object pose estimation dataset for industrial bin-picking (2019). arXiv:1912.12125
49. Correll, N., Bekris, K.E., Berenson, D., Brock, O., Causo, A., Hauser, K., Okada, K., Rodriguez, A., Romano, J.M., Wurman, P.R.: Analysis and observations from the first amazon picking challenge. IEEE Trans. Autom. Sci. Eng. **15**(1), 172–188 (2016)
50. Hackett, D., Pippine, J., Watson, A., Sullivan, C., Pratt, G.: An overview of the darpa autonomous robotic manipulation (arm) program. J. Robot. Soc. Jpn. **31**(4), 326–329 (2013)
51. Morgan, A.S., Hang, K., Bircher, W.G., Alladkani, F.M., Gandhi, A., Calli, B., Dollar, A.M.: Benchmarking cluttered robot pick-and-place manipulation with the box and blocks test. IEEE Robot. Autom. Lett. **5**(2), 454–461 (2019)
52. Xiang, Y., Schmidt, T., Narayanan, V., Fox, D.: Posecnn: A convolutional neural network for 6d object pose estimation in cluttered scenes (2017b). arXiv:1711.00199
53. Kumar, S., Majumder, A., Dutta, S., Raja, R., Jotawar, S., Kumar, A., Soni, M., Raju, V., Kundu, O., Behera, E.H.L. et al.: Design and development of an automated robotic pick and stow system for an e-commerce warehouse (2017). arXiv:1703.02340
54. Sun, Y., Falco, J., Cheng, N., Choi, H.R., Engeberg, E.D., Pollard, N., Roa, M., Xia, Z.: Robotic grasping and manipulation competition: task pool. In: Robotic Grasping and Manipulation Challenge, pp. 1–18. Springer, Berlin (2016)
55. Murase, H., Nayar, S.K.: Visual learning and recognition of 3-D objects from appearance. Int. J. Comput. Vision **14**(1), 5–24 (1995)

56. Hema, C.R., Paulraj, M.P., Nagarajan, R., Sazali, Y.: Segmentation and location computation of bin objects. Int. J. Adv. Robot. Syst. **4**(1), 57–62 (2007)
57. Tombari, F., Stefano, L.D.: Object recognition in 3d scenes with occlusions and clutter by hough voting. In: Proceedings of the 2010 Fourth Pacific-Rim Symposium on Image and Video Technology, PSIVT '10, pp. 349–355 (2010)
58. Liu, M.-Y., Tuzel, O., Veeraraghavan, A., Taguchi, Y., Marks, T.K., Chellappa, R.: Fast object localization and pose estimation in heavy clutter for robotic bin picking. Int. J. Robot. Res. **31**(8), 951–973 (2012)
59. Tiwan, P., Boby, R.A., Roy, S.D., Chaudhury, S., Saha, S.K.: Cylindrical pellet pose estimation in clutter using a single robot mounted camera. In: Proceedings of Conference on Advances In Robotics, pp. 1–6 (2013)
60. Raghuvanshi, T., Chaudhary, S., Jain, V., Agarwal, S., Chaudhury, S.: Automated monocular vision based system for picking textureless objects. In: Proceedings of Conference on Advances In Robotics, pp. 1–6 (2015)
61. Matthias, N., David, D., Dirk, H., Joerg, S., Alexander, B., Jun, L., Reinhard, B.: Mobile bin picking with an anthropomorphic service robot. In: 2013 IEEE International Conference on Robotics and Automation (ICRA), pp. 2327–2334 (2013a)
62. Buchholz, D., Futterlieb, M., Winkelbach, S., Wahl, F.M.: Efficient bin-picking and grasp planning based on depth data. In: 2013 IEEE International Conference on Robotics and Automation (ICRA), pp. 3245–3250 (2013)
63. Domae, Y., Okuda, H., Taguchi, Y., Sumi, K., Hirai, T.: Fast grasp ability evaluation on single depth maps for bin picking with general grippers. In: 2014 IEEE International Conference on Robotics and Automation (ICRA), pp. 1997–2004 (2014)
64. SICK: Gigabit 3D vision for tough environments (2017). https://www.sick.com/us/en/vision/3d-vision/ruler/c/g138562
65. MicroEpsilon: scanCONTROL selection (2017b). http://www.micro-epsilon.in/2D_3D/laser-scanner/model-selection/
66. MicroEpsilon: Laser line triangulation (2017a). http://www.micro-epsilon.in/service/glossar/Laser-Linien-Triangulation.html
67. Zhang, Q., Pless, R.: Extrinsic calibration of a camera and laser range finder (improves camera calibration). In: 2004 IEEE/RSJ International Conference on Intelligent Robots and Systems, 2004. (IROS 2004). Proceedings, vol. 3, pp. 2301–2306 (2004)
68. Unnikrishnan, R., Hebert, M.: Fast extrinsic calibration of a laser rangefinder to a camera. CMU Technical Report (2005)
69. Horaud, R., Dornaika, F.: Hand-eye calibration. Int. J. Robot. Res. **14**(3), 195–210 (1995)
70. Roy, M., Boby, R.A., Chaudhary, S., Chaudhury, S., Roy, S.D., Saha, S.K.: Pose estimation of texture-less cylindrical objects in bin picking using sensor fusion. In: 2016 IEEE/RSJ International Conference on Intelligent Robots and Systems, IROS 2016, pp. 2279–2284 (2016)
71. Haralick, R.M., Shapiro, L.G.: Computer and Robot Vision, vol. 2. Addison-Wesley Publishing Company, Masachusets, USA (1993)
72. Tsai, R.Y.: A versatile camera calibration technique for high-accuracy 3D machine vision metrology using off-the-shelf TV cameras and lenses. IEEE J. Robot. Autom. **3**(4), 323–344 (1987)
73. Zhang, Z.: A flexible new technique for camera calibration. IEEE Trans. Pattern Anal. Mach. Intell. **22**(11), 1330–1334 (2000)
74. Richardson, A., Strom, J., Olson, E.: Aprilcal: Assisted and repeatable camera calibration. In: 2013 IEEE/RSJ International Conference on Intelligent Robots and Systems (IROS), pp. 1814–1821 (2013)
75. Lepetit, V., Moreno-Noguer, F., Fua, P.: Epnp: An accurate o(n) solution to the pnp problem. Int. J. Comput. Vision **81**, 155–166 (2009)
76. Boby, R.A.: Hand-eye calibration using a single image and robotic picking up using images lacking in contrast. In: 2020 International Conference Nonlinearity, Information and Robotics (NIR), pp. 1–6. IEEE (2020)

77. Rusu, R.B.: Semantic 3d object maps for everyday manipulation in human living environments. PhD thesis, TU Munich, Munich, Germany (2009)
78. Pless, R.: Code for calibration of cameras to planar laser range finders (2004). http://research.engineering.wustl.edu/~pless/code/calibZip.zip
79. Matthias, N., David, D., Dirk, H., Joerg, S., Alexander, B., Jun, L., Reinhard, B.: Mobile bin picking with an anthropomorphic service robot. In: 2013 IEEE International Conference on Robotics and Automation (ICRA), pp. 2327–2334 (2013b)
80. Liu, M.-Y., Tuzel, O., Veeraraghavan, A., Taguchi, Y., Marks, T.K., Chellappa., R.: Fast object localization and pose estimation in heavy clutter for robotic bin picking. Int. J. Robot. Res. **31**(8), 951–973 (2012)

Chapter 3
Uncertainty and Sensitivity Analysis

Uncertainty in engineering systems comes from various sources such as manufacturing imprecision, assembly errors, model variation, and stochastic operating conditions. It is important to mathematical model for the spatial measurement uncertainty and the sensitivity analysis of the robots kinematic parameters for a sensor-fused robotic system. Studies on uncertainty analysis were done using probabilistic models without emphasizing the context of multiple sensor mounted on the robot. And mostly the local sensitivity analysis is studied to estimate the variation of input parameters on the output using the Jacobian. This chapter presents the propagation relationship of the uncertainties in the camera's image space, the 2D space of the laser scanner, the Cartesian space in the world coordinate, and the tool space. Since a vision sensor is a vital sensor for localization, it is essential to have performance maps. Here, we perform an analytical and quantitative simulation to model uncertainty in the pose from camera and depth sensors that are mounted on a KUKA KR5 Arc robot's end-effector. A quantitative analysis of the spatial measurement of the sensor-fused system is performed by utilizing the covariance matrix in the depth image space and the calibrated sensor parameters using the proposed mathematical model. Moreover, sensitivity analysis assesses the contributions of the inputs to the uncertainty. The total sensitivity analysis using Sobol's method is presented to assesses the contributions of each input kinematic parameters, i.e., Denvait–Hartenberg (DH) parameters on the uncertainty in the output, i.e., pose inside the robot's workspace.

3.1 Background

The demand for industrial robots with higher absolute positioning accuracy has been growing continuously, especially in aerospace, medical treatment, and robot-assisted measurement, where positions are defined in a virtual space to an absolute or

A. A. Hayat et al., *Vision Based Identification and Force Control of Industrial Robots*,
Studies in Systems, Decision and Control 404,
https://doi.org/10.1007/978-981-16-6990-3_3

relative coordinate system. Based on investigation of the error contribution from various sources, the robot geometric errors are responsible for about 90% of the total positioning error [1]. Calibrations are measurements that determine quantity measurement values delivered by a device under test with those of a calibration standard of known accuracy. Because calibration devices, calibration standards, reagents, and tools are not perfect, environmental conditions, processes, procedures, and people are also imperfect and variable, and each calibration has an uncertainty associated with it [2]. This chapter presents the propagation relationship of the uncertainties in the cameras image space, the 2D space of the laser scanner, the real Cartesian space in the world coordinate, and the tool space, which is the end-effector of the robot with the mapping function between the spaces.

Several sources of errors are involved in the operation of a manipulator, such as mechanical imprecision, poor sensing resolution, unwanted vibration, and structural deformation [3]. For industrial manipulators, [4] reported two types of errors: geometric and non-geometric. Geometric errors fall under the category of mechanical factors such as link length and joint axis error. Differences in manufacturing accuracy limits, assembly misalignment, joint offsets error referred to as geometric errors and accounts for 95% of the total positional error [5]. Similar conclusion were reported in [6]. The effect of joint position error during a precise insertion task using a stochastic model was investigated by [7]. Figure 1.2 in Chap. 1 shows the sources of error in a manipulator. The uncertainties in the knowledge of kinematic parameters such as link lengths, axis vectors, joint angles, and positions become significant when the required precision is comparable to the error caused by imprecision in these parameters. It becomes vital to quantify the effect of individual kinematic parameters on the pose accuracy.

The concept of reliability as a measure of manipulator performance was analyzed in [8]. The first-order sensitivity functions were derived using Jacobian matrix-based technique for position and orientation error study in [9]. This idea of sensitivity is local as it is at a particular pose. The sensitivity of the output y say in $y = f(x)$ w.r.t., x can be obtained using partial derivative at a given point $x = x^*$ as $(\partial y/\partial x)_{x=x^*}$. Figure 3.1 shows the sensitivity of two parameters, and it can be easily concluded that the sensitivity of the first parameter is more. We have used the global or total sensitivity approach first proposed by [10]. This is one of the elegant methods capable of computing Total Sensitivity Index (TSI), which captures the main effect along with the interactions among the parameters.

The kinematic sensitivity was defined as the maximum error that can occur in the Cartesian workspace as a result of bounded uncertainties in the joint space [11]. Two separate sensitivity indices were assigned for the position and rotation of the parallel manipulator in [11]. The sensitivity of the inaccuracies in the link length and payload weight on the mass matrix was investigated in [12]. The focus of this chapter are as following:

- Modeling of the uncertainty in each sensor (camera, laser range sensor, and the manipulator) according to their characteristics.

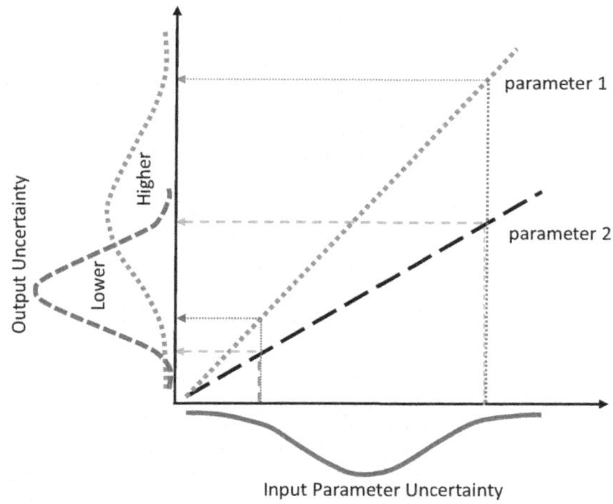

Fig. 3.1 Relation between output and input uncertainty for low- and high-sensitive parameters

- A covariance-based uncertainty propagation relationship for the camera output, laser range scanner output, and the manipulator end-effector position and orientation.
- A suitable selection of parameters to reduce the overall uncertainty in a sensor-fused system.

3.2 Formulation

Sensitivity analysis is defined as the study of how the different sources of input variations can be apportioned quantitatively to the output variations. During the evaluation of a robot, it will be vital to know the relative importance of the kinematic parameters that define the robot's architecture. Figures 3.1 and 3.2 show schematic diagrams to explain the sensitivity analysis.

3.2.1 Local Sensitivity Analysis and Uncertainty Propagation

Consider the set of variable **x**, which is converted to another output variables **y** via the given transformation function as

$$\mathbf{y} = \mathbf{f}(\mathbf{x}) \tag{3.1}$$

$$\mathbf{f}(\mathbf{x}) = [f_1(\mathbf{x}) \;\; f_2(\mathbf{x}) \;\; \cdots \;\; f_M(\mathbf{x})] \tag{3.2}$$

$$\mathbf{x} = [x_1 \; x_2 \; \cdots \; x_N] \tag{3.3}$$

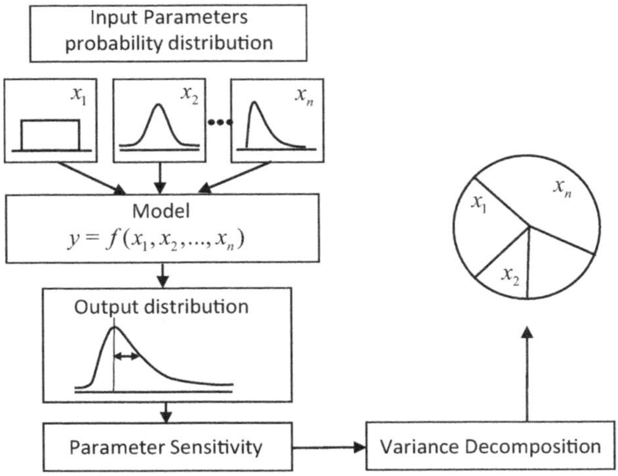

Fig. 3.2 Schematic diagram for sensitivity analysis

If $f_i(\mathbf{x})$ is a linear function of $\mathbf{x}\,(\forall i = 1, 2, \cdots, M)$, then we can represent it as shown below:

$$\mathbf{y} = \mathbf{A}\mathbf{x} \tag{3.4}$$

If $\mathbf{x}$ follows a Gaussian distribution (i.e., $\mathbf{x} \sim \mathcal{N}(\boldsymbol{\mu}_x, \boldsymbol{\Sigma}_x)$), then we can easily find the probability distribution for $\mathbf{y}$ using the following equations:

$$\begin{aligned} \boldsymbol{\mu}_y &= \mathbf{A}\boldsymbol{\mu}_x \\ \boldsymbol{\Sigma}_y &= \mathbf{A}\boldsymbol{\Sigma}_x\mathbf{A}^T \end{aligned} \tag{3.5}$$

Hence, the distribution for the transformed vector $\mathbf{y}$ is given by $\mathcal{N}\left(\boldsymbol{\mu}_y, \boldsymbol{\Sigma}_y\right)$. If the function $\mathbf{f}$ is not linear, then we can linearize it using the Taylor Series expansion, and get the desired distribution for $\mathbf{y}$. Writing Taylor series expansion for $f_i(\mathbf{x})$, as in Eq. (3.6)

$$f_i(\mathbf{x} + \delta\mathbf{x}) = f_i(\mathbf{x}) + \sum_{j=1}^{N} \frac{\partial f_i(\mathbf{x})}{\partial x_j}\delta x_j + \frac{1}{2}\sum_{l=1}^{N}\sum_{m=1}^{N} \frac{\partial^2 f_i(\mathbf{x})}{\partial x_l \partial x_m}\delta x_l \delta x_m + \cdots \tag{3.6}$$

If we rewrite Eq. (3.6) again as Eq. (3.7),

$$f_i(\mathbf{x} + \delta\mathbf{x}) = f_i(\mathbf{x}) + \mathbf{g}^T\delta\mathbf{x} + \frac{1}{2}\delta\mathbf{x}^T\mathbf{H}\delta\mathbf{x} + \varepsilon\left(||\delta\mathbf{x}||^3\right) \tag{3.7}$$

In Eq. (3.7), **g** represents the gradient vector and **H** represents the Hessian matrix.

$$\mathbf{g} = g_f(\mathbf{x}) = \nabla_f(\mathbf{x}) = \frac{d\mathbf{f}(\mathbf{x})}{d\mathbf{x}} = \begin{bmatrix} \frac{\partial \mathbf{f}(\mathbf{x})}{\partial x_1} \\ \vdots \\ \frac{\partial \mathbf{f}(\mathbf{x})}{\partial x_N} \end{bmatrix} \tag{3.8}$$

Using Eq. (3.7), we can obtain Eq. (3.9)

$$\mathbf{f}(\mathbf{x} + \delta\mathbf{x}) \approx \begin{bmatrix} f_1(\mathbf{x}) \\ \vdots \\ f_M(\mathbf{x}) \end{bmatrix} + \begin{bmatrix} \frac{\partial f_1}{\partial x_1} & \cdots & \frac{\partial f_1}{\partial x_N} \\ \vdots & \ddots & \vdots \\ \frac{\partial f_M}{\partial x_1} & \cdots & \frac{\partial f_M}{\partial x_N} \end{bmatrix} \begin{bmatrix} \delta x_1 \\ \vdots \\ \delta x_N \end{bmatrix} \tag{3.9}$$

Rewriting Eq. (3.9) as follows:

$$\mathbf{f}(\mathbf{x} + \delta\mathbf{x}) \approx \mathbf{f}(\mathbf{x}) + \mathbf{J}_f(\mathbf{x})\,\delta\mathbf{x} \tag{3.10}$$

Using Eq. (3.10), we can write the error in **y**i.e.,$\delta\mathbf{y}$ as

$$\delta\mathbf{y} = \mathbf{J}_f \delta\mathbf{x} \tag{3.11}$$

Taking $\delta\mathbf{x}$ to be Normally distributed (i.e., $\delta\mathbf{x} \sim \mathcal{N}(\mathbf{0}, \boldsymbol{\Sigma}_x)$), then the distribution of $\delta\mathbf{y}$ will also be normal distribution with zero mean and covariance given by Eq. (3.12)

$$\boldsymbol{\Sigma}_y = \mathbf{J}_f \boldsymbol{\Sigma}_x \mathbf{J}_f^T \tag{3.12}$$

Using Eqs. (3.5) and (3.12), we can obtain the propagated covariance Matrix depending on the transformation that has been applied.

3.2.1.1 Uncertainty Ellipsoid

Since we are modeling the uncertainties in our system as Gaussian noise, a very easy and robust way of quantifying or verifying the uncertainty model is by analyzing the uncertainty ellipsoid. For zero-mean error [13], we can write the distribution as follows:

$$\mathscr{P}_{a_\varepsilon}(\mathbf{A}_\varepsilon) = \frac{1}{|2\pi|^{\frac{K}{2}} |\boldsymbol{\Lambda}_\varepsilon|^{\frac{1}{2}}} \exp\left(-\frac{1}{2}\mathbf{A}_\varepsilon^T \boldsymbol{\Lambda}_\varepsilon^{-1} \mathbf{A}_\varepsilon\right) \tag{3.13}$$

On the basis of above probability distribution, we can see the equation for the equal height contours is given by

$$\mathbf{A}_\varepsilon^T \boldsymbol{\Lambda}_\varepsilon^{-1} \mathbf{A}_\varepsilon = C^2 \tag{3.14}$$

Equation (3.14) represents the equation for an ellipsoid , hence using this equation, we can find the contour with the certain probability. The probability with which the error vector will lie within this ellipsoid will be given by

$$\mathscr{P} = 1 - \frac{K}{(2)^{\frac{K}{2}} \Gamma\left(\frac{K}{2}+1\right)} \int_{C}^{\infty} X^{K-1} e^{-X^2/2} \mathrm{d}X \tag{3.15}$$

So, we can obtain the value of C on the basis of certain confidence level (95%, 99%, etc.) and the degree of freedom. Using the combination of degree of freedom and confidence level, we can obtain C from chi-square table as it follows a chi-squared distribution.

3.2.2 *Global Sensitivity Analysis Based on Variance*

Sensitivity analysis based on variance rely on the uncertainties of the input parameters. With this technique, the importance of a given input factor can be quantified which is termed as sensitivity index. It reflects the fractional contribution to the variance in the output due to uncertainties in the input.

Let a non-linear scalar mathematical model be represented as $y = f(\mathbf{x})$, where $\mathbf{x} \equiv [x_1, \ldots, x_n]^T$ is the vector of input parameters and y be the corresponding scalar output. The contribution of uncertainty (represented by probability density function) in the input parameter x_i to the output y can be quantified as [14]

$$S_i = \frac{\mathrm{Var}(\mathbb{E}[y|x_i])}{\mathrm{Var}(y)} \tag{3.16}$$

where 'Var' stands for the variance and ($\mathbb{E}[y|x_i]$) is the conditional expectation of y given x_i. The measure S_i is referred as first-order sensitivity index (FSI). This is similar to the partial derivative ($\partial y/\partial x_i$) concept, which is known as local sensitivity. The FSI measures the contribution of each input parameters on the variance of the output. It does not consider the combined effect of interaction of two parameters on the output variance.

The sensitivity measure based on variance presented in [10] is widely used for total sensitivity analysis (TSA). This technique measures the main effect of a given parameter and its associated influences. For example, let there be three input parameters, x_1, x_2, and x_3, then the total effect of parameter x_1 on the output is

$$TS(x_1) = S_1 + S_{12} + S_{13} + S_{123} \tag{3.17}$$

where $TS(x_i)$ is the total sensitivity index (TSI) for parameter x_i. The term S_1 represents the first-order sensitivity of parameter x_i. By first order, one means the fractional contribution of a single parameter to the output variance. The term S_{ij}

denotes the second-order sensitivity which is an indicator of the measure of the fractional contribution of parameter interactions to the output variance. Similarly, one can also define third-order sensitivity denoted by S_{ijk}, and so on.

3.2.2.1 Computation of Total Sensitivity Index

The central idea behind the Sobol's approach for computing the sensitivity index is the decomposition of the function $f(\mathbf{x})$ into the summands of increasing dimensionality as

$$\begin{aligned} f(\mathbf{x}) = & f_0 + f_1(x_1) + \cdots + f_n(x_n) + f_{12}(x_1, x_2) + f_{23}(x_2, x_3) + \\ & \cdots + f_{123}(x_1, x_2, x_3) + f_{234}(x_2, x_3, x_4) + \cdots \end{aligned} \tag{3.18}$$

where $\mathbf{x}$ is a vector of n uncertain inputs $x_1, x_2, \cdots, x_n$. It is assumed here that the inputs are independently and uniformly distributed within a hypercube, i.e., $x_i \in [0, 1], \forall i = 1, \cdots, n$. This is without any loss of generality because any input can be transformed onto the hypercube. The above expression in general can be written as

$$f(\mathbf{x}) = f_0 + \sum_{i}^{n} f_i(x_i) + \sum_{i<j}^{n} f_{ij}(x_i, x_j) + \ldots + f_{1\ldots n}(x_1, \ldots, x_n) \tag{3.19}$$

$$\equiv f_0 + \sum_{s=1}^{n} \sum_{i_1<\ldots<i_s}^{n} f_{i_1,\ldots,i_s}(x_{i_1}, \ldots, x_{i_s}) \tag{3.20}$$

where $1 \le i_1 < \cdots < i_s \le n$ and $s = 1, \cdots, n$. The integer s is termed as the *order* or the *dimension* of the index. For example, i_1 means first-order terms like $\{x_1, x_2, \ldots\}$, i_2 means second-order term like $\{(x_1, x_2), (x_2, x_3), \cdots\}$, and so on, and finally $x_{i_s} = x_1, x_2, \ldots, x_n$. The total number of summands in the above equation is equal to 2^n. For Eq. 3.19 to hold true f_0 must be constant and the integral of every summand over any of its own variable must be zero as

$$f_0 = \text{constant} \tag{3.21}$$

$$\int f_{i_1,\ldots,i_s}(x_{i_1}, \ldots, x_{i_s}) dx_k = 0 \;\; \forall \, k = i_1, \ldots, i_s \tag{3.22}$$

Note that in Eq. (3.22), $\int f(\mathbf{x}) d\mathbf{x}$ means multidimensional integrals $(\int_0^1 \ldots \int_0^1)$ over all the differential variables present in an interval [0, 1]. The consequence of the above two conditions, i.e., Eqs. 3.21 and 3.22 is that each summands in Eq. 3.18 are orthogonal to one other, i.e.,

$$\int f_{1\ldots i} f_{j\ldots n} \mathrm{d}x = 0 \tag{3.23}$$

and the other consequence showed by [10] is that the decomposition in Eq. 3.20 is unique and all the terms can be evaluated using multidimensional integrals, as

$$f_0 = \int f(\mathbf{x})\mathrm{d}\mathbf{x} \tag{3.24}$$

$$f_i(x_i) = -f_0 + \int f(\mathbf{x}) \prod_{k \neq i} dx_k \equiv -f_0 + \int_0^1 \cdots \int_0^1 f(\mathbf{x}) d\mathbf{x}_{\sim i} \tag{3.25}$$

$$f_{ij}(x_i, x_j) = -f_0 - f_i(x_i) - f_j(x_j) + \int f(\mathbf{x}) \prod_{k \neq i,j} \mathrm{d}x_k \tag{3.26}$$

$$\equiv -f_0 - f_i(x_i) - f_j(x_j) + \int_0^1 \cdots \int_0^1 f(\mathbf{x}) \mathrm{d}\mathbf{x}_{\sim ij} \tag{3.27}$$

The above expression indicates that the integration is done over all the parameters except x_i in Eq. 3.25 and except x_{ij} in Eq. 3.26. The symbol $\sim$ is used to denote "complementary of". Similarly, the form can be extended for other higher terms.

The variance-based sensitivity can be deduced from the above equations. By definition, the total variance is *mean of square minus square of mean*. Squaring Eq. 3.20 and integrating it over I^n will give

$$D = \int f^2(\mathbf{x}) d\mathbf{x} - f_0^2 \tag{3.28}$$

$$\equiv \sum_{i}^{n} \sum_{i_1 < \ldots < i_s}^{n} \int f^2_{i_1,\ldots,i_s} \mathrm{d}x_{i_1}, \ldots, \mathrm{d}x_{i_s} \tag{3.29}$$

D is constant and is called variance of $f(\mathbf{x})$. The partial variance of each input variables term is computed from the terms:

$$D_{i_1,\ldots,i_s} = \sum_{i}^{n} \sum_{i_1 < \cdots < i_s}^{n} \int f^2_{i_1,\ldots,i_s} \mathrm{d}x_{i_1}, \ldots, \mathrm{d}x_{i_s} \tag{3.30}$$

The global sensitivity indices (SI) [15] is then defined as

$$S_{i_1,\ldots,i_s} = \frac{D_{i_1,\ldots,i_s}}{D} \tag{3.31}$$

where S_i is the first-order SI for factor x_i, on the output, i.e., partial contribution of x_i on variance of $f(\mathbf{x})$. Similarly, S_{ij} for $i \neq j$ is the second-order SI which gives the effect of interactions between x_i and x_j and so on. The total sensitivity index can be written as follows:

$$TS_i = S_i + \sum_{i<j} S_{ij} + \sum_{j \neq i, k \neq i, j<k} S_{ijk} + \cdots \tag{3.32}$$

To estimate the first-order and total sensitivity indices, [15] developed the Monte Carlo-based method. To evaluate the total sensitivity index (TSI) of the parameters, all the integral terms in summand need not be calculated. With only single integral calculation, the TSI can be computed. For example, with three input parameters in Eq. (3.17), the TSI for parameter x_1 can be computed by considering two subset of the variables as $u = x_1$ and $v = (x_2, x_3)$ as

$$TS(x_1) = S_1 + S_{1,2} + S_{1,3} + S_{1,2,3} = 1 - S_{2,3} \equiv 1 - S_{\sim 1}$$

$$\text{or,} \qquad TS(u) = 1 - S_v \tag{3.33}$$

So, the variables are partitioned into two subsets, one containing variable x_i and other containing its complementary, i.e., the rest of the variables. Similarly, grouping can be applied for estimating $TS(\cdot)$ of other model parameters. The Monte Carlo implementation of the above equations were done in Matlab for the simulation.

3.3 Illustration with a KUKA KR5 Arc Robot

This section presents the implementation of the technique presented in Sect. 3.2.2.1 for evaluating the sensitivity of the kinematic parameters of the industrial manipulator KUKA KR5 Arc. The architecture of the robot is shown in Fig. 3.3b. DH parameters with their identified values are listed in Table 3.1. The net transformation matrix to obtain the position and orientation of the end-effector (EE) is given by

$$\mathbf{T} = \prod_{i=1}^{6} \mathbf{T}_i \tag{3.34}$$

where $\mathbf{T}$ is the 4×4 homogenous transformation matrix. The X, Y, and Z-positions of the EE of the robots in world frame is given by the first, second, and third elements in the fourth column of the homogenous matrix $\mathbf{T}$. While the Euler angles α_Z, β_Y and γ_X, i.e., orientation about Z-, Y-, and X-axis, respectively, were obtained from the 3×3 rotation matrix in the first three rows and column of $\mathbf{T}$. They are explained in any text book on robotics, say in [16]. In short, the pose of the robot as a function of its DH parameters (b_i, a_i, α_i) and joint variables (θ_i) is given by

$$[X, Y, Z, \alpha_z, \beta_y, \gamma_x]^T_{6\times 1} = FK_k(b_i, a_i, \alpha_i, \theta_i) \tag{3.35}$$

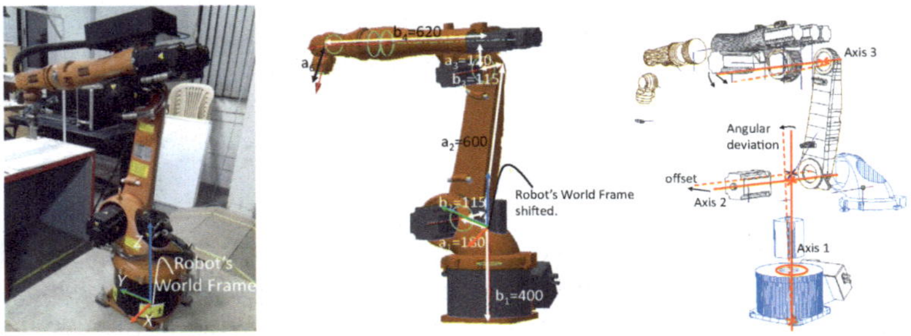

a) Installed robot b) DH parameters c) Misalignment in axes and positions

Fig. 3.3 Error causing perturbation in DH parameters

Table 3.1 DH parameters of KUKA KR5 Arc robot from specification sheet

Joint	b (mm)	θ (degrees)	a (mm)	α (degrees)
1	400.00	0	180.00	90.00
2	0	90.00	600.00	180.00
3	0	0	120.00	90.00
4	620.00	180.00	0	90.00
5	0	–90.00	0	90.00
6	0	90.00	0	0

where FK_k is a highly non-linear function with $k = 1, 2, \ldots, 6$ for the six functions corresponding to the six output parameters, i.e., X, Y, Z, α_z, β_y, and γ_x, respectively. The DH parameters with subscript i are used here for the number of links. For six links, the forward kinematics model is a function of $(6 \times 4) = 24$ DH parameters. Reasons for using the DH parameters for the SA are (1) They are most widely used, and (2) The major cause for the variation in these parameters can be visualized with the misalignment (Fig. 3.3c) in the following fashion:

- The variation in twist angle α can occur during the assembly process. Mainly due to improper alignment of the actuator shaft with the hub. This is diagrammatically shown in Fig. 3.4b.
- The variation in the link length and joint offset can occur because of the manufacturing and finishing processes of making the particular link. The clearance in the joint angles can also be one of the reason. The combined effect with offset and angular misalignment is shown in Fig. 3.4c.
- Lastly, the variation in joint angles θ_i's can occur because of the gain and offset in the joint angle position sensors. The slack in timing belt of the transmission system, joint stiffness, and backlash in the gear also contribute to the variation in joint angles.

  In view of the above, facts the SA will help in inferring the major factors contributing to the variation in the pose or performance. For studying the effect of the

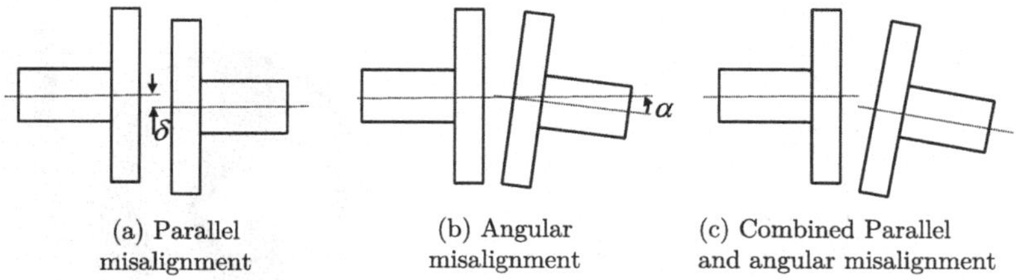

Fig. 3.4 Types of misalignment causing error

Table 3.2 Joint limits of KUKA KR5 Arc

	θ_1	θ_2	θ_3	θ_4	θ_5	θ_6
Minimum	−155	−65	−68	−350	−130	−350
Maximum	155	180	105	350	130	350

parameter sensitivity, the robot shown in Fig. 3.3 was considered. The errors or misalignments which can occur in KUKA KR5 Arc robot on the first three axes is shown in Fig. 3.3c. Note that any other mechanisms can be taken into account and the method presented here can be followed for its SA.

Assumptions which were considered in carrying out the sensitivity analysis are as follows:

(i) The deviation is assumed to exist in all the DH parameters. Here, the deviation error means the difference from the parameters in the CAD or in the nominal parameter.
(ii) Normal distribution was assumed for the variation in the parameters. The mean of the distribution was taken as the nominal value of the parameters listed in Table 3.1. It is briefly explained in Appendix A.2.
(iii) For linear or length parameters, i.e., b_i and a_i, the variations in the parameters were assumed with standard deviation (SD), $\sigma_l = 1$ mm, whereas for angular parameter, α_i was taken as $\sigma_a = 1°$. As a result, the interval in which these parameters range is $[b_i - 3\sigma_l, b_i + 3\sigma_l]$, $[a_i - 3\sigma_l, a_i + 3\sigma_l]$, $[\alpha_i - 3\sigma_a, \alpha_i + 3\sigma_a]$. The perturbation in joint angle θ_i was selected on the basis of type of analysis desired, i.e., whether it was for fixed pose or in the whole workspace.
(iv) The workspace of the robot shown in Fig. 3.5a was created using the limits on each joint listed in Table 3.2.

In this study, the effect of the uncertainty in the linear and angular parameters of DH parameters on the position and orientation errors of KUKA KR5 Arc robot is presented for two cases:

- **Case 1**: For a given pose inside, the workspace which was selected near a good manipulability zone as shown in Fig. 3.5b. The six joint values for the pose selected were [0°, 70°, 12°, 0°, 0°, 0°]. The perturbations in the length parameters b_i and

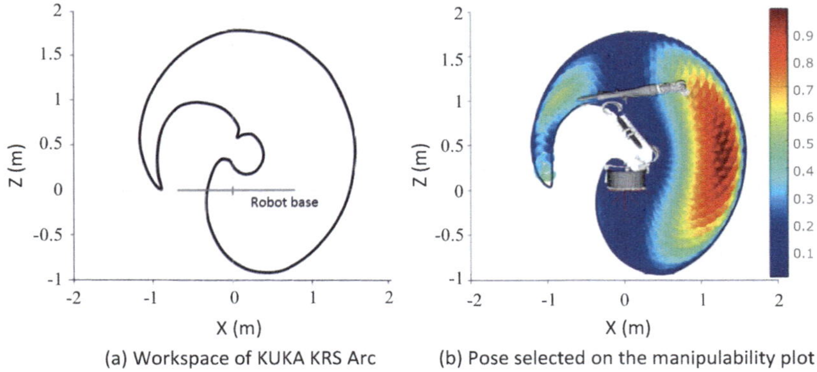

Fig. 3.5 Workspace of the KUKA KR5 Arc robot with its manipulability plot

a_i, and the twist angle α_i were kept same as mentioned in steps mentioned in Algorithm 3, whereas the variation in the joint angle, θ_i was taken as $1°$.

- **Case 2**: The analysis was extended to the whole workspace. For this, the range of joint angles θ_i's were considered as the joint limits of the KUKA KR5 Arc mentioned in Table 3.2.

Algorithm 3 Steps for estimating sensitivity index of the DH parameters

1: Select the number of input parameters for sensitivity analysis. Here, DH parameters of the KUKA KR5 Arc were considered.
2: Assign the interval bound for the perturbation values of these parameters. Gaussian distribution was taken to assign the perturbation each DH parameters. Intervals were taken as $(b_i \pm \delta b)$, $(a_i \pm \delta a)$ for length parameters and $(\alpha_i \pm \delta\alpha)$, $(\theta_i \pm \delta\theta)$ for angular parameters.
3: The sensitivity analyses were performed for the parameters selected on its outputs, namely, position and orientation of the robot. They were calculated using Eq. (3.34).
4: Sensitivity indices of the parameters were then found using the Sobols method.

The algorithm for calculating the Total Sensitivity Index (TSI) of DH parameters were written in MATLAB® and tested on Intel® Core, i7-3770 CPU and 3.40 GHz with 8 GB RAM, 64 bit Windows 7 Operating System.

3.3.1 Results and Discussions

This section presents the results obtained after the sensitivity analyses and the discussion on it for the two cases mentioned.

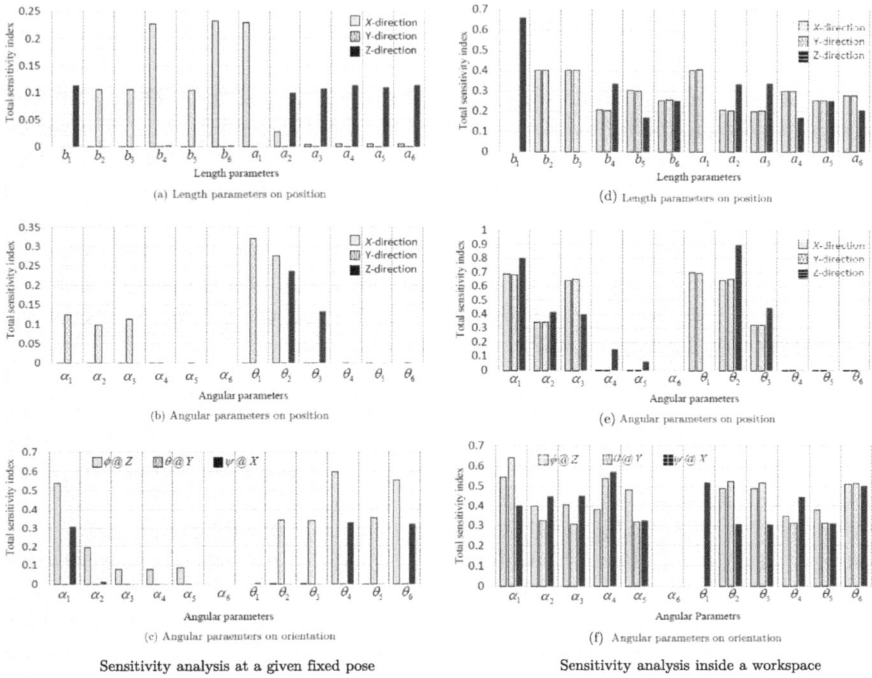

Fig. 3.6 Local and total or global sensitivity to variation in DH parameters

For Case 1—At a given pose

The results plotted in the Fig. 3.6a–c, based on which the following conclusions are drawn:

- The sensitivity of the DH parameters at a fixed pose of the robot is shown in Fig. 3.6, it can be concluded that the length parameters only affect the positioning of the robot, i.e., motion along the $X-$, Y-, and Z-directions not the rotations about $X-$, Y-, and Z-axes.
- From the bar charts of Fig. 3.6a, it can safely concluded that the X-direction is affected most by the parameters $b_{4,6}$ and a_1.
- The angular parameters in Fig. 3.6b depict that the twist angles of the first three joints contribute to the variation in the Y-direction. The perturbations in first three joint angles about the initial pose, i.e., $[0°, 70°, 12°, 0°, 0°, 0°]$ only contribute to the position.
- It is also evident from the sensitivity analyses on orientation, Fig. 3.6b, that only the angular parameters contribute. The last three joint variables, i.e., $\theta_{4,5,6}$, affect the rotations parameter about X and Y represented by Euler angles α_z, γ_y, respectively.

For Case 2—Complete workspace

- The bar charts in Fig. 3.6d–f show the results for the sensitivity of DH parameters corresponding to the full work envelope of the robot. From Fig. 3.6a, d, it can be concluded that in both the figures TSI of b_1 is only reflected in Z-direction. This is geometrically true by the inherent definition of joint offset and is also validated using one at a time (OAT) analysis shown for b_1 in Fig. 3.7. Note that the scale of the contribution increases from 0.12 (Fig. 3.6a to 0.68 (Fig. 3.6d). The variations in the parameter b_1 only shifts the position of the EE in the Z-direction and hence not a major concerned for the designer. The base frame can also be attached at the intersection of the axes 1 and 2 as shown in Fig. 3.3b.
- Similar to Fig. 3.6a, d shows the effect of b_2, b_3 in the Y-direction with the equal SI in the X-direction. This is true for the complete workspace analysis as with the change of pose of EE the direction of joint offset parameters $b_{2,3}$ also points along the X-axis of the base frame. The same TSI in both X and Y-direction also reflects the correctness of the identified index.
- From the bar charts in Fig. 3.6d, e, it can be concluded that the parameters $b_{(4,5,6)}$, $a_{(1,2,\cdots,6)}$, and $\alpha_{(1,2,3)}$ have TSI in all the X-, Y-, and Z-directions. These are also the critical parameters which contribute to the positioning accuracy. One can calculate a combined index along three directions by taking the norm of the three TSIs of the individual parameters. The advantage of the separate TSI for three directions and

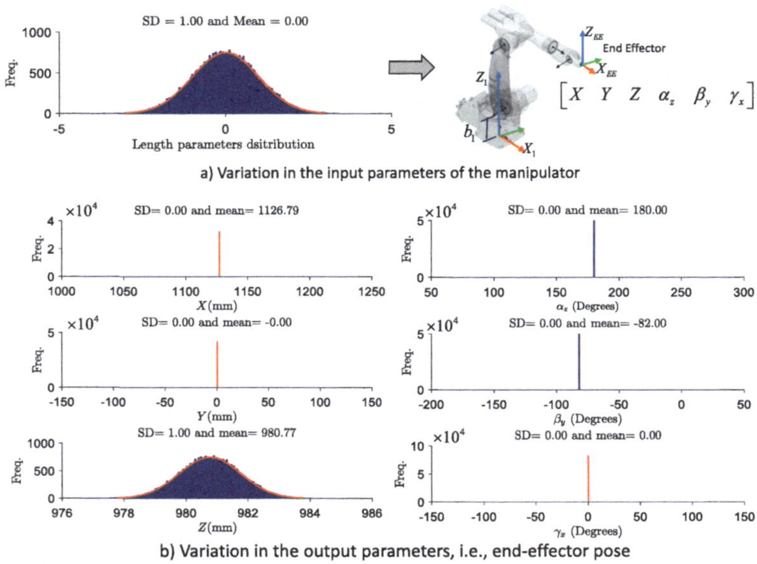

Fig. 3.7 Local sensitivity on pose by varying a single parameter

orientations is that they provide individual picture. This may help in the calibration process of the robot as well by selecting the high sensitivity parameters.

- In Fig. 3.6f, the TSI for orientation is shown. It can be concluded that all the angular parameters, i.e., $\alpha_{1,\ldots,5}$ and $\theta_{1,\ldots,6}$ contribute to the orientation. The assembly between the joints 1 and 2 govern the magnitude of α_1 which has high TSI in orientation and position. Similarly, $\alpha_{2,3}$ equally contribute to the error in the pose. Hence, the most critical among all constant parameters are $\alpha_{1,2,3}$ which is depicted in Fig. 3.6e, f.

The TSI analysis gave an estimate of the relative importance of the input factors on the model output. The results presented in the form of bar graph were verified using a simple approach, i.e., by evaluating the changes in the model outputs with respect to the perturbation in a single parameter one at a time (OAT) also termed as local sensitivity. Figure 3.7 shows the effect of the variation in the joint offset of firstlink b_1 on the pose ($[X, Y, Z, \alpha_z, \beta_y, \gamma_x]$). The variation in b_1 was assigned with SD of 1 mm. The net result of this variation was reflected only in the Z-coordinate of the EE. This can be geometrically visualized as well that any change in the dimension of joint offset b_1 (Fig. 3.3a) will occur only in the Z-direction. Hence, only one output, i.e., along the Z-direction will vary. The results obtained from perturbation of OAT parameters were listed in Table 3.3. The TSI found for the DH parameters gave a system performance indicator in the presence of kinematic data variability with a combination of input parameters. The uncertainty in the parameters are inevitable, and as a result, this method was presented to quantify it by a metric assigned to the DH parameters based on the sensitivity. This information was utilized for the experimental calibration of the manipulator, as explained next.

Table 3.3 Model outputs standard deviation (SD) using OAT

Parameters	SD (X)	SD (Y)	SD (Z)	SD(γ_x)	SD (β_y)	SD (α_z)
b_1	0	0	0.9996	0	0	0
b_2	0	0.9996	0	0	0	0
b_3	0	0.9996	0	0	0	0
a_1	0.9996	0	0	0	0	0
a_2	0.3419	0	0.9394	0	0	0
a_3	0.1391	0	0.9899	0	0	0
α_1	0	0.0101	0	179.996	0	0.0071
α_2	0	0.0121	0	179.9997	0	0.0015
α_3	0	0.0128	0	180.0009	0	0

3.4 Uncertainty in Sensor Fused System

Spatial uncertainty in the system can be caused due to several reasons such as measurement error, lightning error, image pixel quantization error, robots kinematic parameters etc. Such uncertainties of the system are dealt in this section of the chapter as shown in Fig. 3.8. In this chapter, we model the uncertainty in the sensor fused system as shown in Fig. 3.8. Therefore, we need to consider the effect of the uncertainty in the robot's frame as well. This is achieved by considering the kinematic identification of the robot's geometric parameters and its effect on the sensors mounted on the end effector. It is followed by the calibration steps to find the transformation between the end-effector's coordinate frame, camera coordinate frame, and the laser scanner coordinate frame. Both of them can be combined to obtain the transformation between camera and laser sensor. Kinematic calibration of the robot influences the accuracy of all the processes utilizing robot kinematics particularly the relation between the robot and the end-effector coordinate frames.

Transformation from the grid coordinate system to the EE coordinate system as shown in Fig. 3.9 which is given by $\mathbf{T}_G^E$ and can be mathematically formulated as follows:

3.4.1 *Uncertainty in Robot*

Let $\mathbf{X}_{EE}$ be the point in robot End-effector coordinate system $\{EE\}$ and $\mathbf{X}_G$ be a point in frame $\{G\}$. Then these are related as below:

$$\mathbf{x}_{EE} = \mathbf{T}_G^{EE}\,\mathbf{x}_G \tag{3.36}$$

Where $\mathbf{T}_G^{EE}$ is the transformation from frame $\{G\}$ to $\{EE\}$ as shown in Fig. 3.9

$$\mathbf{T}_G^{EE} = \begin{bmatrix} \mathbf{Q} & \mathbf{t} \\ 0 & 1 \end{bmatrix} \tag{3.37}$$

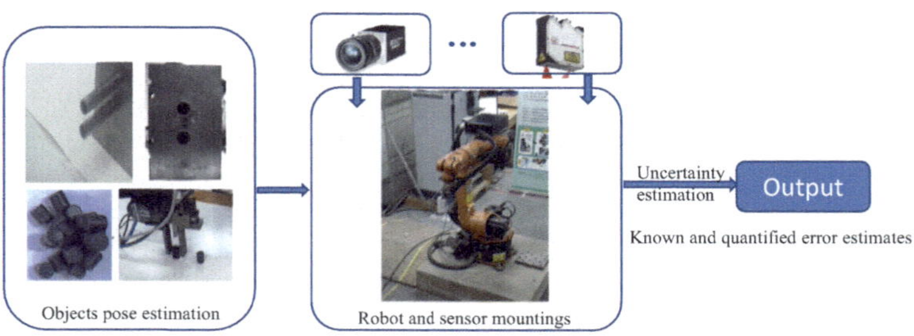

Fig. 3.8 Effect of various uncertainties on the final pose estimates

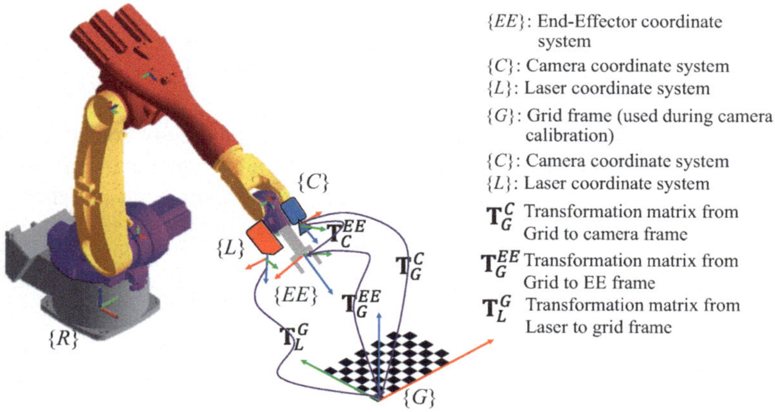

Fig. 3.9 Sensor fused robotic bin-picking system with the different transformation matrices among the coordinate frames assigned

$$\mathbf{x}_{EE} = \mathbf{Q}\mathbf{x}_G + \mathbf{t} \tag{3.38}$$

Here, $\mathbf{Q}$ is 3×3 rotational matrix and $\mathbf{t}$ is 3×1 translational vector, in $\{EE\}$. Due to some mechanical sources of errors like gear backlash, elasticity, robot forward kinematics, etc., noise is introduced in $\mathbf{T}_G^{EE}$ defined in Eq. (3.37). Which can be represented as

$$\hat{\mathbf{T}}_G^E = \begin{bmatrix} \mathbf{Q} & \mathbf{t} + \Delta\mathbf{t} \\ 0 & 1 \end{bmatrix} \tag{3.39}$$

which shows that, in a calibrated system transformation matrix, we need to consider its effect, to get accurate pose estimate at the robot end-effector, whereas an analysis of the measured values of orientation errors and its effect on the rotation matrix is studied as: The errors in orientation be represented by $\delta\alpha$, $\delta\beta$ and $\delta\gamma$. The change in orientation matrix is $\delta\mathbf{Q}$ which is represented by

$$\delta\mathbf{Q} \equiv \begin{bmatrix} c(\delta\beta)c(\delta\alpha) - 1 & s(\delta\gamma)s(\delta\alpha)c(\delta\gamma)s(\delta\beta)c(\delta\alpha) & -c(\delta\gamma)s(\delta\alpha) - s(\delta\beta)s(\delta\gamma)c(\delta\alpha) \\ c(\delta\beta)s(\delta\alpha) & -s(\delta\gamma)c(\delta\alpha) - c(\delta\gamma)s(\delta\beta)s(\delta\alpha) & c(\delta\gamma)c(\delta\alpha)s(\delta\beta)s(\delta\gamma)s(\delta\alpha) - 1 \\ s(\delta\beta) & c(\delta\beta)c(\delta\gamma) - 1 & c(\delta\beta)s(\delta\gamma) \end{bmatrix} \tag{3.40}$$

Note that s and c represent sin and cos, respectively. Subsequently, for small values of α, β and γ, the following can be assumed:

$$\lim_{\alpha\to 0, \beta\to 0, \gamma\to 0} \delta\mathbf{Q} \equiv \begin{bmatrix} 0 & -\delta\beta & -\delta\alpha \\ \delta\alpha & -\delta\gamma & 0 \\ \delta\beta & 0 & \delta\gamma \end{bmatrix} \tag{3.41}$$

For average values of repeatability (corresponding to 10% speed) from,

$$\delta \mathbf{Q} \equiv \begin{bmatrix} 0 & -0.0009 & -0.0006 \\ 0.0006 & -0.0003 & 0 \\ 0.0009 & 0 & 0.0003 \end{bmatrix} \tag{3.42}$$

Since the values are negligible, the errors in orientation are not considered while computing **Q**. This shows that, in a calibrated system transformation matrix, we need to consider its effect to get accurate pose estimate at the robot's end-effector.

$$\hat{\mathbf{x}}_G = \mathbf{x}_G + \Delta \mathbf{x}_G \tag{3.43}$$

where $\Delta \mathbf{x}_G$ is $\mathcal{N}(\mu, \sigma^2)$

$$\hat{\mathbf{x}}_{EE} = \mathbf{T}_G^{EE} \, \hat{\mathbf{x}}_G \tag{3.44}$$

and assuming

$$\hat{\mathbf{x}}_{EE} = \mathbf{x}_{EE} + \Delta \mathbf{x}_{EE}$$

from Eq.(3.44), we have

$$\hat{\mathbf{x}}_{EE} = \begin{bmatrix} \mathbf{Q} & \mathbf{t} + \Delta \mathbf{t} \\ 0 & 1 \end{bmatrix} \begin{bmatrix} \mathbf{x}_G + \Delta \mathbf{x}_G \\ 1 \end{bmatrix} \tag{3.45}$$

$$\mathbf{x}_{EE} + \Delta x_{EE} = \begin{bmatrix} \mathbf{Q}(\mathbf{x}_G + \Delta \mathbf{x}_G) + \mathbf{t} + \Delta \mathbf{t} \\ 1 \end{bmatrix} \tag{3.46}$$

we have

$$\mathbf{x}_{\mathbf{EE}} = \begin{bmatrix} \mathbf{Q}\mathbf{x}_G + \mathbf{t} \\ 1 \end{bmatrix}$$

$$\Delta \mathbf{x}_{EE} = \begin{bmatrix} \mathbf{Q} \Delta \mathbf{x}_G + \Delta \mathbf{t} \\ 0 \end{bmatrix} \tag{3.47}$$

$$\Delta \mathbf{x}_{EE} = \mathbf{Q} \Delta \mathbf{x}_G + \Delta \mathbf{t} \tag{3.48}$$

$$\|\Delta \mathbf{x}_{EE}\| \leq \|\Delta \mathbf{x}_G\| + \|\Delta t\| \tag{3.49}$$

3.4.2 Uncertainty in Camera

Uncertainty in the camera matrix has to be modeled because matrix transforms the 3D world coordinate system to 2D image coordinate as described in the Eq. (3.50). Error in calibration matrix is measurement error that occurs due to manual measurements and sensor geometry.

$$\lambda \mathbf{p} = \mathbf{M}\mathbf{x} \tag{3.50}$$

Image formation is represented by

$$\mathbf{p} = \begin{bmatrix} u \\ v \\ 1 \end{bmatrix}, \mathbf{M} = \begin{bmatrix} m_{11} & m_{12} & m_{13} & m_{14} \\ m_{21} & m_{22} & m_{23} & m_{24} \\ m_{31} & m_{32} & m_{33} & m_{34} \end{bmatrix}, \text{and } \mathbf{x} = \begin{bmatrix} x \\ y \\ z \\ 1 \end{bmatrix} \tag{3.51}$$

Camera projection matrix (CPM) **M** is estimated using least square method. Here, u and v represent image coordinates and λ is the scaling factor. Grid frame $\{G\}$ are represented by $\mathbf{x}_w$. camera matrix M represents the general 3D to 2D transformation. Uncertainty in the CPM has an impact on the back projected to $\{G\}$. It is assumed that the corner points are co-planar. In the setup shown in Fig. 3.9, matrix will remain fixed for the entire operation of pick and place as, image is always taken from the same fixed pose. The CPM plays an important role during sensor fusion and efficient bin picking process. Therefore, error modeling in the back projection matrix becomes important.

Another important limitation imposed with just 3D segmentation can be seen in Fig. 2.16(g). Whenever multiple closely placed pellets are present 3D segmentation fails and image segmentation can be used to find out the centroid of those pellets using CPM by means of backward projection. The estimate of pose can then be refined. The backward projection is finding in frame $\{G\}$ from its image coordinates and CPM were found. Centroid of the pellets in 3D in frame $\{G\}$ are estimated by back projecting image data into world data. So, if the CPM is erroneous then the estimated centroids of the pellets are also erroneous, thus effecting the pick-up pose.

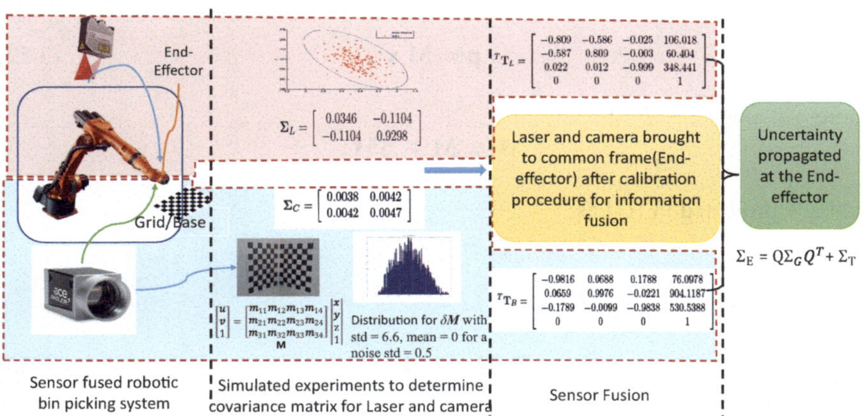

Fig. 3.10 Image demonstrating the importance of sensor fusion

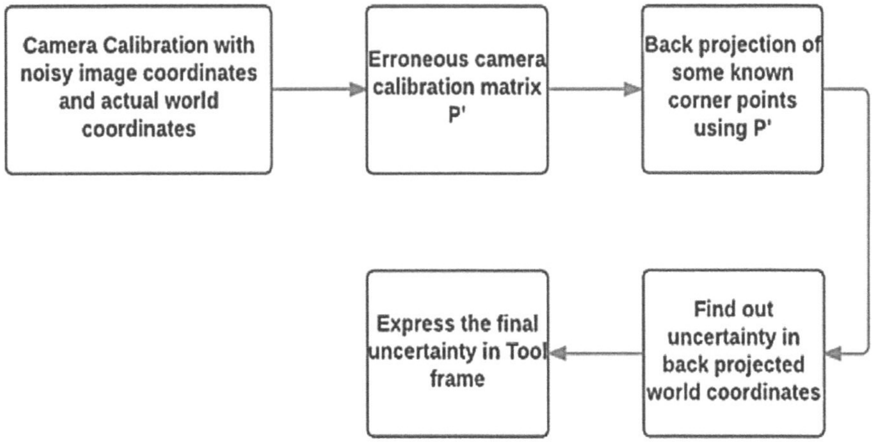

Fig. 3.11 Flow diagram for estimating the uncertainty propagated to EE

In this section we will find the effect of erroneous CPM on back projected $\{G\}$s, and model its uncertainty to justify the accuracy with which camera parameters should be estimated so that the back projected $\{G\}$ are within an acceptable error tolerance. Simulation procedure is shown in Fig. 3.11. To find out the uncertainty back projected in frame $\{G\}$ due to erroneous CPM, we generate erroneous CPM from the least square method using noisy image coordinates and actual coordinates in frame $\{G\}$ of known corner points of the checker board shown in Fig. 3.12.

In order to get the noisy image coordinates, we generated the Gaussian noises with zero mean and different standard deviations ranging from 0.05 pixel to 1 pixel and added it to the actual image coordinates. So we get different erroneous camera calibration matrix after calibration for different noise standard deviations.

$$\lambda \mathbf{p} = \mathbf{M}^{'} \mathbf{x}^{'} \tag{3.52}$$

where

$$\mathbf{M}^{'} = \mathbf{M} + \Delta \mathbf{M}$$

and back projected in $\{G\}$ as:

$$\mathbf{x}^{'} = \mathbf{x} + \Delta \mathbf{x}$$

from Eqs. 3.50 and 3.52, we get

$$\mathbf{M}\mathbf{x} = \mathbf{M}^{'} \mathbf{x}^{'} \tag{3.53}$$

$$\mathbf{M}\mathbf{x} = (\mathbf{M} + \Delta \mathbf{M})\,(\mathbf{x} + \Delta \mathbf{p})$$

$$\Delta \mathbf{x} = \mathbf{M}^{+}\,(\Delta \mathbf{M} x) \tag{3.54}$$

We † is the Moore–Penrose inverse function. The error in CPM is $\Delta\mathbf{M}$ is Gaussian. Let $\Delta(M)$ be independent with zero mean and variance $\sigma^2_{\Delta\mathbf{M}}$, then

$$f_{\Delta\mathbf{M}}(\Delta\mathbf{m}) = \left(\frac{1}{\sqrt{(2\pi\sigma^2_{\Delta M})}} \exp^{-\frac{\Delta\mathbf{M}\delta\mathbf{M}^T}{\sigma_{\Delta\mathbf{M}^2}}} \right) \tag{3.55}$$

from Eq. 3.54, using the uncertainty definition and model presented in [17], we have

$$f_{\Delta\mathbf{x}}(\Delta\mathbf{x}) = \left(\frac{1}{\sqrt{\left(2\pi \left(MM^T\right)^{-1} xx^T\sigma^2_{\Delta M}\right)}} \right) exp^{\left(\frac{\Delta x \Delta x^T}{2\left(MM^T\right)^{-1} xx^T\sigma^2_{\Delta M}}\right)} \tag{3.56}$$

Therefore, if error in the image coordinates $[\Delta u, \Delta v]$ is Gaussian and the error distribution in camera projection matrix (CPM) is $\Delta\mathbf{M}$ is also Gaussian. Hence, according to the Eq. (3.56) it is proved that error back projected to the $\{G\}$ frame will also be Gaussian.

3.4.3 Uncertainty in Laser Sensor

After performing the calibration procedure, we obtain the transformation matrix from laser sensor to end-effector. We transform all the points in the laser frame to one common base which can be obtained by the following transformation:

$$\mathbf{T}_L^G = \mathbf{T}_{EE}^G\, \mathbf{T}_L^{EE} \tag{3.57}$$

Note that $\mathbf{T}_L^G$ represents the transformation from laser $\{L\}$ to $\{G\}$ which is obtained by multiplying $\mathbf{T}_{EE}^G$ and $\mathbf{T}_L^{EE}$. $\mathbf{T}_{EE}^G$ is the transformation from $\{EE\}$ to $\{G\}$ which can be obtained using the transformation from robot to $\{G\}$. Additionally, $\mathbf{T}_L^{EE}$ is the transformation from laser to $\{EE\}$ obtained after calibration. We represent a point in the laser frame as $\mathbf{x}_L$ and a point in the $\{EE\}$ as $\mathbf{x}_G$, which are related as follows:

$$\mathbf{x}_G = \mathbf{T}_L^G\, \mathbf{x}_L \tag{3.58}$$

It may be interpreted that the Y-coordinate for the point in Laser Frame is zero, because the laser sensor gives the measurement along its profile which is in XZ plane only as already discussed previously while performing calibration. The laser coordinates $\mathbf{x}_L$ is then given as

$$\mathbf{x}_L = \begin{bmatrix} x_L \\ 0 \\ y_L \\ 1 \end{bmatrix} \tag{3.59}$$

It is now possible to find the covariance matrix of the uncertainty that got propagated to $\{EE\}$ (Fig. 3.9) and further to $\{G\}$ from laser frame because of the uncertainty in observation by laser sensor. The covariance matrix of the uncertainty in laser frame $\{L\}$ (Fig. 3.9) calculated experimentally as:

$$\Sigma_L = \begin{bmatrix} 0.0346 & -0.1104 \\ -0.1104 & 0.9298 \end{bmatrix} \tag{3.60}$$

The above matrix will be used to estimate the propagated covariance matrix in next section.

3.4.4 Uncertainty Propagation to $\{EE\}$

Propagation of uncertainty from the $\{G\}$ to the $\{EE\}$ (see Fig. 3.9). Let Σ_G is the covariance matrix estimated in the $\{G\}$ after sensor calibration and similarly, Σ_t is the covariance matrix estimated in the translation vector due to robot error as explained in Sect. 3.4.1. therefore, from Eq. (3.38) uncertainty in the $\{EE\}$ can be given as

$$\Sigma_{\mathbf{EE}} = \mathbf{Q}\Sigma_G\mathbf{Q}^T + \Sigma_{\mathbf{t}} \tag{3.61}$$

where $\Sigma_{\mathbf{EE}}$ is the final uncertainty at the end-effector of the Robot.

For clarity the steps used to determine the uncertainty propagation is discussed in Appendix A.2.

3.4.4.1 Uncertainty Propagated from Camera to $\{EE\}$

Matrix $\mathbf{T}_C^G$ represents the transformation from camera to grid, i.e., nothing but back projection. Note that $\mathbf{T}_{EE}^G$ is the transformation from end-effector to $\{G\}$ and $\mathbf{T}_C^{EE}$ is the transformation from camera to $\{EE\}$ frame obtained after calibration.

$$\mathbf{T}_C^G = \mathbf{T}_{EE}^G\,\mathbf{T}_C^{EE} \tag{3.62}$$

We represent a point in the camera frame as $\mathbf{x}_C$ and a point in frame $\{G\}$ as $\mathbf{x}_G$, which are related as follows:

$$\mathbf{x}_G = \mathbf{T}_C^G\,\mathbf{x}_C \tag{3.63}$$

After performing the camera calibration procedure and uncertainty modeling, we obtain the transformation matrix from camera to end-effector. We transform all the points in the camera frame to one common base. For illustration purpose consider uncertainty calculation in the EE frame for the corner point [17.96, 0, 19.6] with a noise standard deviation 0.5 is mathematically formulated as follows: Covariance matrix in base frame is given as:

$$\Sigma_G = \begin{bmatrix} 0.0038 & 0.0042 \\ 0.0042 & 0.0047 \end{bmatrix} \tag{3.64}$$

Grid to EE transformation (Fig. 3.9) is given as:

$$\mathbf{T}_G^{EE} = \begin{bmatrix} -0.9816 & 0.0688 & 0.1788 & 76.0978 \\ 0.0659 & 0.9976 & -0.0221 & 904.1187 \\ -0.1789 & -0.0099 & -0.9838 & 530.5388 \\ 0 & 0 & 0 & 1 \end{bmatrix} \tag{3.65}$$

and robot uncertainty covariance matrix is

$$\Sigma_t = .0001\mathbf{I} \tag{3.66}$$

Where $\mathbf{I}$ is 4×4 is identity matrix. From Eq. (3.61), final uncertainty in the $\{EE\}$ due to camera is given as:

$$\Sigma_{E_{camera}} = \begin{bmatrix} 0.0025 & -0.0001 & -0.0039 \\ -0.0001 & 0.0001 & -0.0002 \\ 0.0039 & -0.0002 & 0.0065 \end{bmatrix} \tag{3.67}$$

3.4.4.2 Uncertainty Propagated from Laser Sensor to $\{EE\}$

Similarly, uncertainty in the laser frame will be propagated to the $\{EE\}$ frame and final uncertainty at the end-effector due to laser using Eqs. (3.60) and (3.61) is given by:

$$\Sigma_{E_{Laser}} = \begin{bmatrix} 0.0187 & 0.0146 & -0.0663 \\ 0.0146 & 0.0115 & -0.024 \\ -0.0663 & -0.024 & 0.09341 \end{bmatrix} \tag{3.68}$$

3.4.5 *Final Uncertainty in the Sensor Fused System*

This section discusses the final uncertainty in the sensor fused picking system. In the Sect. 3.4.2, we described the uncertainty only due to camera; similarly, in the Sect.

3.4.3, we discussed uncertainty due to laser only. Taking cue from these sections, when we try to get the pose of object using just 3D information, we take into consideration the laser uncertainty as shown in Fig. 3.12. But when the confidence of the pose is low from this information, 2D information takes the lead and pose of the object is determined by fusing both the information shown in Fig. 3.12. Therefore, final uncertainty of the system is given by Eq. (3.69).

$$\Sigma_{Fused} = \begin{bmatrix} -0.2133 & 0.0129 & -0.568 \\ 0.0129 & 0.0045 & -0.2152 \\ -0.568 & -0.2152 & 0.8689 \end{bmatrix} \times 10^{-3} \tag{3.69}$$

3.5 Experimental Results

For uncertainty analysis of the sensor-fused system, a camera and a 2D laser scanner mounted on the end-effector of a KUKA KR5 industrial robot are used. The laser scanner used is Micro-epsilon ScanCONTROL 2900–100 with 0.12 μm resolution [18]. The scanner has ability to provide 2D information namely, the depth and distance of a pixel representing the measured object along the line of laser projection [19]. Meanwhile, the camera used is Basler Pilot, piA2400-17gc with 2454 pixels × 2056 pixels resolution. Conversion of the 2D data from the laser scanner to a single coordinate frame (B) is crucial for bin-picking applications. The calibration of the laser scanner to the camera is also important to enable the verification of the data emerging from any of the sensors. This will decrease errors in pose estimation

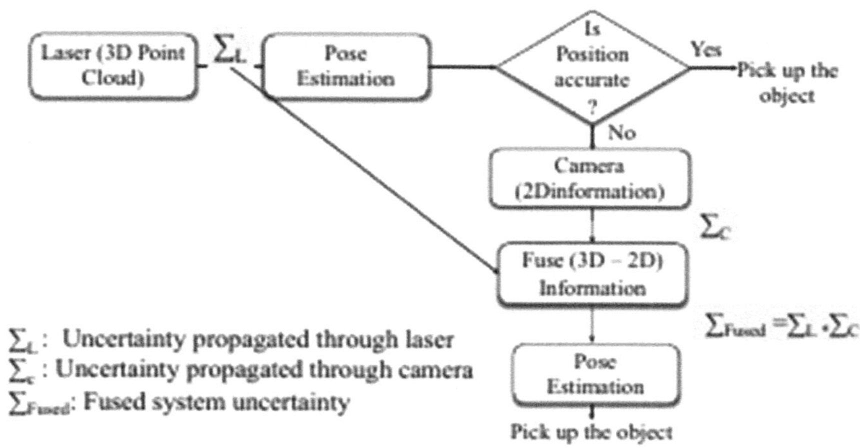

Fig. 3.12 Final uncertainty in the sensor-fused picking system

and also will quicken the process of picking up from the bin by quickly detecting the disturbance that may have happened after each pick-up.

3.5.1 *Effect of Uncertainty on Camera Calibration Matrix*

Known points from the calibration grid shown in Fig. 3.13 are used to find out the uncertainty in back projected world coordinates due to erroneous calibration matrix. Later using erroneous calibration matrix and known image coordinates, we have found out the back projection error in world coordinates using Eq. (3.56). It is clear that the error in back projection also follows Gaussian distribution as shown in Table 3.4. These results help in determining the accuracy with which the Image coordinates need to be made for the calibration process so that the back-projected world coordinates are within an acceptable error tolerance to determine the pick up point on the cylindrical pellets.

The uncertainty in back projected world coordinate is calculated and visualized for a single corner point with different noise standard deviation and for different corner points with fixed noise standard deviation as shown in Fig. 3.14 (Table 3.5).

3.5.2 *Effect of Uncertainty in Laser Frame*

The distribution of the uncertainty in laser frame is depicted as Gaussian Curve in Fig. A.4. The uncertainty ellipsoid for the uncertainty in the laser frame with respect to the robot tool/end-effector is depicted in Fig. 3.17, which also shows the observations in laser frame which fits well in the observed ellipsoid. Laser sensor used in our setup works on the principle of triangulation as shown in Fig. 3.15 [19].

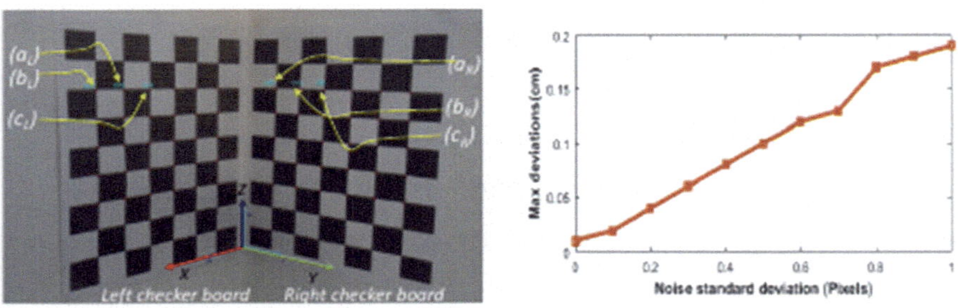

a) Two checker boards patterns b) Plot for standard deviation

Fig. 3.13 a) 3D Checker board used for camera calibration (size of each square is 12 mm), b) Uncertainity for a single corner point in world frame with different noise standard deviations

Table 3.4 Uncertainty ellipses for different corner points on checker board. Size of each square is 12 mm. We are moving left to right on a checker board and it is observed that as we move more towards the center of the checker board deviation is less as shown in Fig. 3.12

Test case 	u	x_G	Σ'_G	$\sqrt{\|\Sigma'_G\|_2}$ (max deviation)	Uncertainty ellipse unit in cm
a_L	$\begin{bmatrix} 402 \\ 422.2 \end{bmatrix}$	$\begin{bmatrix} 17.96 \\ 0 \\ 19.6 \end{bmatrix}$	$\begin{bmatrix} 0.0038 & 0.0042 \\ 0.0042 & 0.0047 \end{bmatrix}$	0.0921	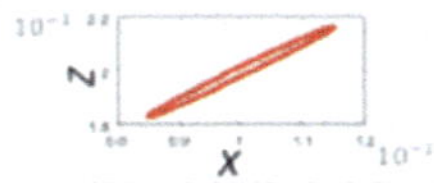
b_L	$\begin{bmatrix} 517.7 \\ 416.8 \end{bmatrix}$	$\begin{bmatrix} 14.90 \\ 0 \\ 19.6 \end{bmatrix}$	$\begin{bmatrix} 0.0027 & 0.0035 \\ 0.0035 & 0.0046 \end{bmatrix}$	0.0853	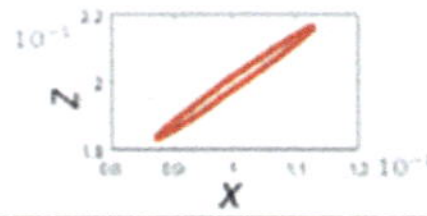
c_L	$\begin{bmatrix} 633.3 \\ 416.8 \end{bmatrix}$	$\begin{bmatrix} 12.09 \\ 0 \\ 19.6 \end{bmatrix}$	$\begin{bmatrix} 0.0018 & 0.0028 \\ 0.0028 & 0.0045 \end{bmatrix}$	0.0793	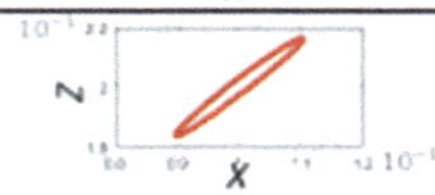
a_R	$\begin{bmatrix} 1170 \\ 419.5 \end{bmatrix}$	$\begin{bmatrix} 0 \\ 3.7 \\ 19.6 \end{bmatrix}$	$\begin{bmatrix} 0.0002 & 0.0008 \\ 0.0008 & 0.0039 \end{bmatrix}$	0.0634	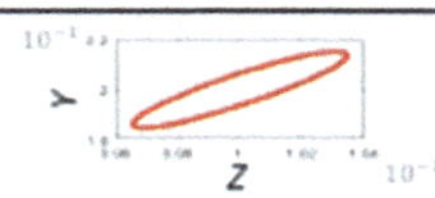
b_R	$\begin{bmatrix} 1271.2 \\ 422.2 \end{bmatrix}$	$\begin{bmatrix} 0 \\ 6.5 \\ 19.6 \end{bmatrix}$	$\begin{bmatrix} 0.0005 & 0.0014 \\ 0.0014 & 0.0040 \end{bmatrix}$	0.067	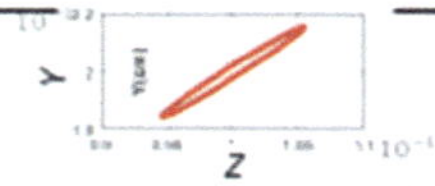
c_R	$\begin{bmatrix} 1372.4 \\ 427.6 \end{bmatrix}$	$\begin{bmatrix} 0 \\ 9.3 \\ 19.6 \end{bmatrix}$	$\begin{bmatrix} 0.0010 & 0.0020 \\ 0.0020 & 0.0041 \end{bmatrix}$	0.0715	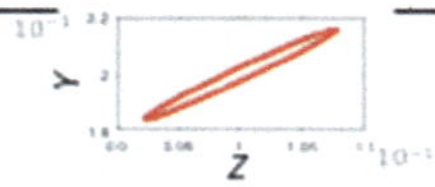

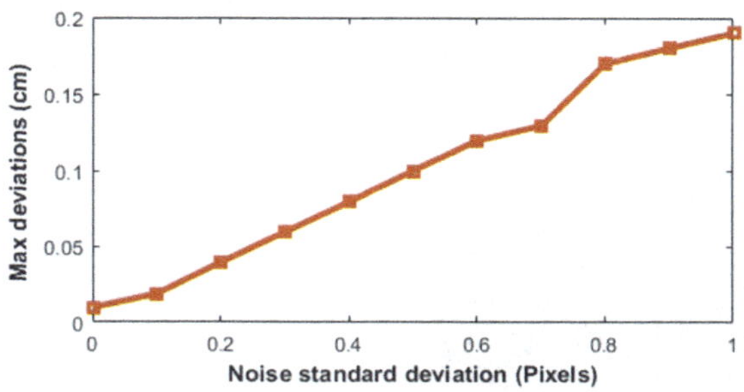

Fig. 3.14 Uncertainty for a single corner point in world frame with different noise standard deviations

Table 3.5 Standard deviation in the positions X and Z for for single corner point with different noise deviations

Noise (σ_n)	X (σ_X) (mm)	Z (σ_Z) (mm)
0.05	0.071	0.063
0.1	0.141	0.126
0.5	0.70	0.62
1	1.40	1.26

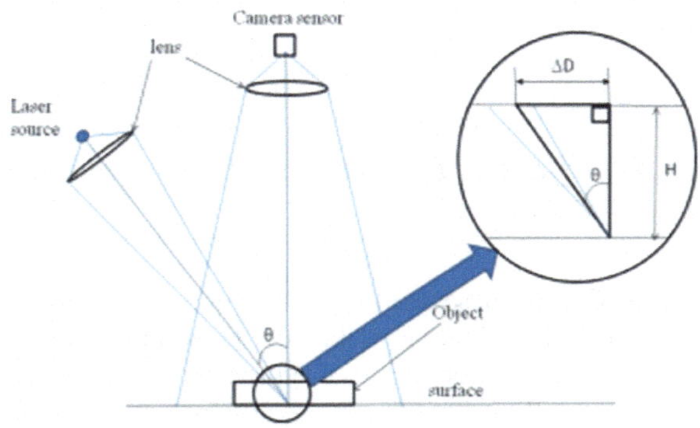

Fig. 3.15 Triangulation principle

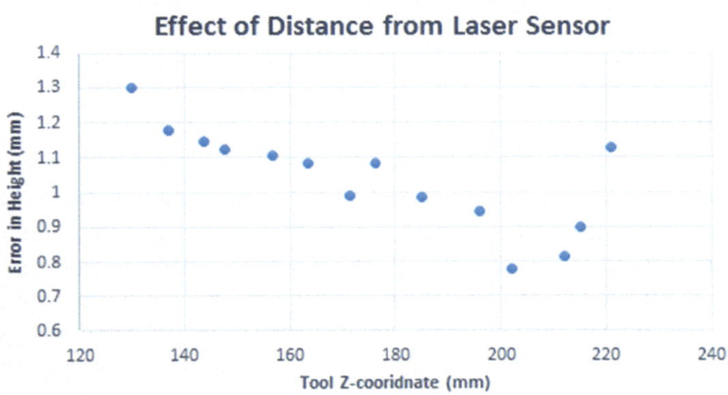

Fig. 3.16 Effect of distance from laser sensor

The reflected laser is sensed by camera lens placed at certain angle to projection axis. Using the geometrical relationships the height of object is calculated.

While performing the experiments using laser sensor, it has been observed that as we bring the sensor close to the measurement area the error in the measurement Z-coordinate of the pointed object initially reduces, but starts increasing after a limit. The error at the various height of the tool from the base is shown in Fig. 3.16. As obtained experimentally, the tool should be placed at a height of approximately 185–200 mm as the error in this region is less than 1 mm.

3.5.3 *Effect of Speed of Robot*

The comparison between the norms of uncertainty matrix obtained in laser frame when the robotic manipulator is run at two different speeds is shown in Table 3.6

We can clearly see that uncertainty increases drastically on increasing the speed. This can be directly related to the fact that calculation of Z-coordinate by laser sensor is done by using the principle of triangulation. When we increase the speed of robotic manipulator, positions sensed by the laser is not correct as it changes its position very quickly and thus the estimate of angle required for triangulation is prone to more errors.

3.5.4 *Effect of Uncertainty Propagated to End-Effector/Tool Frame*

This section discusses the final uncertainty propagation to the end-effector/tool frame, from where the object is picked up. Uncertainty ellipsoids related to the covariance matrix mention in the Eq. (3.61) this are shown in Table 3.7. Figure 3.17 shows the uncertainty ellipsoid of the uncertainty propagated to the tool frame from laser frame and all the points in the Tool Frame falls within that ellipsoid.

3.6 Summary

This chapter proposes a mathematical model for the spatial measurement uncertainty and the total sensitivity analysis of the robots kinematic parameters for a sensor-fused robotic system. In this chapter, we show how the various camera calibration parameters are affected by noisy image measurements. We have established that the error in the image center is Gaussian, and it increases as we move away from the center. Therefore, an increase in the error directly affects the back-projected world coordinates and finally affects the object's pose. It is also important to consider the effect of the orientation matrix to get the accurate pose estimation at the robots end-

Table 3.6 Effect of speed of end-effector

Speed of robotic manipulator (%)	Uncertainty norm
10	0.9811
30	0.9432
50	3.4487
75	4.9951
100	5.9899

Table 3.7 Uncertainty ellipsoid for a single corner point in the EE frame with different noise standard deviation validates the uncertainity ellipsoid equation mentioned in section 3.2.1.1 by following the results as Gaussian distribution

Test case 	u	x_G	Σ'_G	$\sqrt{\|\Sigma_G\|_2}$ (max deviation)	Uncertainty ellipsoid in EE frame unit in cm
a_L	$\begin{bmatrix} 402 \\ 422.2 \end{bmatrix}$	$\begin{bmatrix} 17.96 \\ 0 \\ 19.6 \end{bmatrix}$	$\begin{bmatrix} 0.0024 & -0.0001 & 0.0039 \\ -0.0001 & 0.0000 & -0.0002 \\ 0.0039 & -0.0002 & 0.0064 \end{bmatrix}$	0.0938	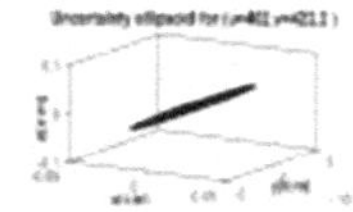
b_L	$\begin{bmatrix} 517.7 \\ 416.8 \end{bmatrix}$	$\begin{bmatrix} 14.90 \\ 0 \\ 19.6 \end{bmatrix}$	$\begin{bmatrix} 0.0015 & -0.0001 & 0.0029 \\ -0.0001 & 0.0000 & -0.0001 \\ 0.0029 & -0.0001 & 0.0058 \end{bmatrix}$	0.0853	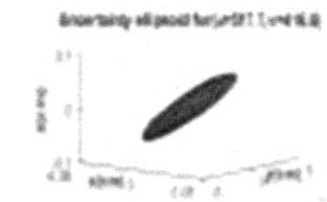
c_L	$\begin{bmatrix} 633.3 \\ 416.8 \end{bmatrix}$	$\begin{bmatrix} 12.09 \\ 0 \\ 19.6 \end{bmatrix}$	$\begin{bmatrix} 0.0009 & -0.0000 & 0.0021 \\ -0.0000 & 0.0000 & -0.0001 \\ 0.0021 & -0.0001 & 0.0054 \end{bmatrix}$	0.0791	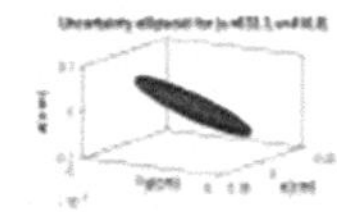
a_R	$\begin{bmatrix} 1170 \\ 419.5 \end{bmatrix}$	$\begin{bmatrix} 0 \\ 3.7 \\ 19.6 \end{bmatrix}$	$\begin{bmatrix} 0.0001 & 0.0001 & -0.0007 \\ 0.0001 & 0.0002 & -0.0007 \\ -0.0007 & -0.0007 & 0.0038 \end{bmatrix}$	0.0638	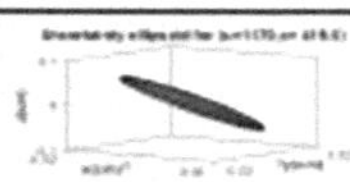
b_R	$\begin{bmatrix} 1271.2 \\ 422.2 \end{bmatrix}$	$\begin{bmatrix} 0 \\ 6.5 \\ 19.6 \end{bmatrix}$	$\begin{bmatrix} 0.0002 & 0.0003 & -0.0008 \\ 0.0003 & 0.0004 & -0.0013 \\ -0.0008 & -0.0013 & 0.0039 \end{bmatrix}$	0.067	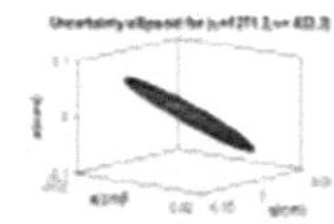
c_R	$\begin{bmatrix} 1372.4 \\ 427.6 \end{bmatrix}$	$\begin{bmatrix} 0 \\ 9.3 \\ 19.6 \end{bmatrix}$	$\begin{bmatrix} 0.0002 & 0.0004 & -0.0009 \\ 0.0004 & 0.0009 & -0.0019 \\ -0.0009 & -0.0019 & 0.0040 \end{bmatrix}$	0.0713	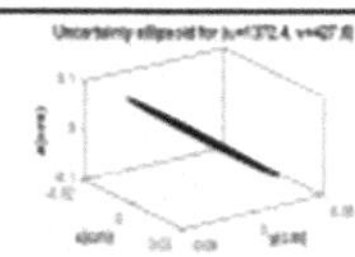

effector. This analysis has practical application and can justify the accuracy with which the measurements need to be made so that the camera calibration parameters are within an acceptable error tolerance. It was also deduced that the laser scanner has to be operated at an optimum height and speed to get the best estimates for the pose. The cumulative effect of the errors is also represented. This spatial uncertainty model was verified by comparing the uncertainty ellipsoids for spatial covariance matrices and was found to be in acceptable error tolerances. The sensitivity analysis helps measure the system's performance by computing the TSI of DH parameters in the presence of kinematic data variability with the combination of input parameters variance using Sobols sensitivity analysis. The uncertainty in the parameters are inevitable and as a result this chapter presented a method to quantify it by a metric

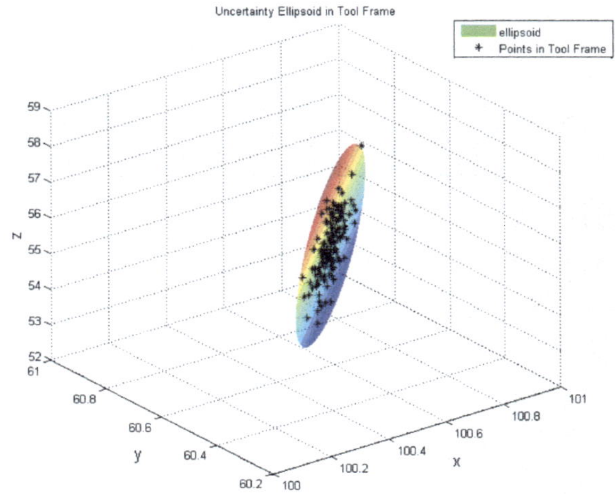

Fig. 3.17 Error ellipsoid for the uncertainty in the tool frame

assigned to the DH parameters based on the sensitivity. This can help in building a bin-picking system using probabilistic approaches.

References

1. Elatta, A., Gen, L.P., Zhi, F.L., Daoyuan, Y., Fei, L.: An overview of robot calibration. Inf. Technol. J. **3**(1), 74–78 (2004)
2. A2LA: G104–guide for estimation of measurement uncertainty in testing (2014)
3. Mooring, B.W., Roth, Z.S., Driels, M.R.: Fundamentals of Manipulator Calibration. Wiley, New York (1991)
4. Jang, J.H., Kim, S.H., Kwak, Y.K.: Calibration of geometric and non-geometric errors of an industrial robot. Robotica **19**(03), 311–321 (2001)
5. Judd, R.P., Knasinski, A.B.: A technique to calibrate industrial robots with experimental verification. IEEE Trans. Robot. Autom. **6**(1), 20–30 (1990)
6. Stone, H.W.: Kinematic Modeling, Identification, and Control of Robotic Manipulators. Springer, Berlin (1987)
7. Azadivar, F.: The effect of joint position errors of industrial robots on their performance in manufacturing operations. IEEE J. Robot. Autom. **3**(2), 109–114 (1987)
8. Bhatti, P., Rao, S.S.: Reliability analysis of robot manipulators. J. Mech. Transm. Autom. Des. **110**(2), 175–181 (1988)
9. Alasty, A., Abedi, H.: Kinematic and dynamic sensitivity analysis of a three-axis rotary table. In: IEEE Conference on Control Applications, vol. 2, pp. 1147–1152. IEEE (2003)
10. Sobol, I.M.: Sensitivity estimates for nonlinear mathematical models. Math. Model. Comput. Exp. **1**(4), 407–414 (1993)
11. Saadatzi, M.H., Masouleh, M.T., Taghirad, H.D., Gosselin, C., Cardou, P.: Geometric analysis of the kinematic sensitivity of planar parallel mechanisms. Trans. Can. Soc. Mech. Eng. **35**(4), 477–490 (2011)

12. Fukuda, T., Arakawa, A.: Optimal control and sensitivity analysis of two links flexible arm with three degrees of freedom. In: Proceedings of the 28th IEEE Conference on Decision and Control, pp. 2101–2106. IEEE (1989)
13. Van Trees, H.L.: Detection, Estimation, and Modulation Theory. Wiley, New York (2004)
14. Saltelli, A., Ratto, M., Andres, T., Campolongo, F., Cariboni, J., Gatelli, D., Saisana, M., Tarantola, S.: Global Sensitivity Analysis: The Primer. Wiley, New York (2008)
15. Sobol, I.M.: Global sensitivity indices for nonlinear mathematical models and their monte carlo estimates. Math. Comput. Simul. **55**(1), 271–280 (2001)
16. Saha, S.K.: Introduction to Robotics, 2nd edn. Tata McGraw-Hill Education (2014b)
17. Uusitalo, L., Lehikoinen, A., Helle, I., Myrberg, K.: An overview of methods to evaluate uncertainty of deterministic models in decision support. Environ. Model. Softw. **63**, 24–31 (2015)
18. MicroEpsilon: Laser line triangulation (2017a). http://www.micro-epsilon.in/service/glossar/Laser-Linien-Triangulation.html
19. MicroEpsilon: scanCONTROL selection (2017b). http://www.micro-epsilon.in/2D_3D/laser-scanner/model-selection/
20. Boby, R.A.: Vision based techniques for kinematic identification of an industrial robot and other devices. PhD thesis (2018)
21. Kopparapu, S., Corke, P.: The effect of noise on camera calibration parameters. Graph. Model. **63**(5), 277–303 (2001)

Chapter 4
Identification

The robotic system considered in this monograph is a serial manipulator and its performance is directly influenced by system dynamics. The parameters defining the dynamics are dependent on both the kinematic parameters, namely, the DH parameters and dynamic parameters, namely, mass, moment of inertia, and inertia. Estimation of these parameters are the focus of this chapter and is important for the control and simulation. The inverse dynamic modeling using the concept of the Decouple Natural Orthogonal Matrices (DeNOC) matrices were adopted for the identification of base inertial parameters. Important aspects like kinematic parameter, kinematic chain identification, modeling, joint friction, trajectory, and identifiability of the robot parameters are discussed.

4.1 Background

4.1.1 Kinematic Identification

In this section, the literature related to the kinematic identification and dynamic identification are discussed. Generally, the kinematic model inside the robot's controller is assumed same as in the design stage. Practically, it is not true because of the errors due to manufacturing, assembly, etc. Hence, identifying the kinematic parameters after the assembly of the robot is essential to feed in the respective controller.

Note that, [1] DH parameters are the most widely used notation to describe a robot architecture. The need to find the DH parameters from the CAD file of the designed

Supplementary Information The online version contains supplementary material available at https://doi.org/10.1007/978-981-16-6990-3_4.

A. A. Hayat et al., *Vision Based Identification and Force Control of Industrial Robots*, Studies in Systems, Decision and Control 404,
https://doi.org/10.1007/978-981-16-6990-3_4

robot stems from the requirement to predict its errors in positioning [2] and implement a correction model in its offline programming [3]. In a way, offline calibration is suggested for an improved performance of a new robot. On the other hand, calibration of a real robot involves four major steps, namely, modeling, measurements, identification, and correction. The purpose is again to improve the accuracy of the robot with or without changing its geometrical parameters in the controller [4]. All calibration methods require nominal parameters [5, 6]. Additionally, the nominal parameters are used for the operation of the robot in immersive environments [7] and using external devices like haptic devices, exoskeleton [8]. Calibration is also carried for the parallel manipulators [9, 10]. Pose estimation and calibration for precision positioning is presented in [11, 12]. It is thus essential to find the nominal DH parameters expeditiously. Besides calibration is needed which is an efficient method to improve the accuracy of the robot without changing the physical properties of the robot.

In [13], only first three link parameters were estimated using a larger set of measured position data acquired by rotating each link by 180°. In another work by [2], spherically mounted reflectors (SMRs) were attached to each link in order to extract the joint axes of the links using laser tracker (LT). Here, a lesser number of points were captured per joint, but the data points were scattered in a larger workspace. The geometric method was used to identify the kinematic parameters of the legged mobile robot named Tarantula in [14]. Recently, the kinematic identification using vision and LT were discussed in [15] with calibration explained in [16]. The unified approach for the parameter identification of manipulators are discussed in [17]. In this chapter, we intend to identify DH parameters of an industrial robot by limited actuation of each link in a smaller workspace. This will facilitate the calibration of a robot in factory setup where there exists restriction on robot's movement. Generally, accuracy is required in a small workspace of the robot where accurate tasks such as peg-in-tube insertion, assembly, etc., are performed. Hence, calibration in that workspace will be used with the data acquired for identification in that workspace.

4.1.2 Dynamic Identification

Dynamic modeling is concerned with the derivation of Equations of Motion (EOM) of a system at hand. Several methods for formulating EOM were given after the comprehensive formulation of Newton–Euler, for example, d'Alembert's principle, Euler–Lagrange principle, Gibbs–Appell approach, and Kane's method. All the methods to obtain the EOM are essentially equivalent. However, methods differ mainly in the operational simplicity, i.e., ease of derivation either by hand or using a computer.

There are formulations proposed in the literature where advantages of both the NE and EL methods were expressed. The main objective was to formulate a way to annihilate the constraint forces in order to avoid extra calculations of constraints forces with the NE method. The matrix computer-oriented method of satisfying constraint equations to reduce the dimension of NE equations was first introduced by [18].

This method for deriving the EOM annihilates the non-working constraint forces and moments which were determined in the recursive Newton–Euler formulation used by [19]. The efficient method using velocity transformation for dynamic analysis of mechanical systems were reported in [20]. Primarily, the linear and angular velocity of links are related with joint velocities by velocity transformation matrix (VTM). The advantage of VTM is that its columns are orthogonal to the vector of constraint moments and forces. More precisely, column space of VTM is an orthogonal complement to the null space of the transpose of the VTM. Orthogonal complement was coined in [21] for modeling non-holonomic systems. Numerical schemes are available for calculating the velocity transformation matrix using singular value decomposition [22] or QR decomposition [23].

Alternatively, [24] derived an orthogonal complement from velocity constraints naturally without any complex numerical computations and named as a natural orthogonal complement (NOC). This NOC combined with the full-descriptive form of NE equations of motions, led to the minimal set of dynamic EOM by eliminating the constraint forces. The effectiveness of the method was shown in [25] for non-holonomic systems, and in [26] for flexible body systems. NOC was decoupled later [27] into two block matrices, namely, a lower block triangular and a block diagonal, which is referred as Decoupled NOC (DeNOC). The recursive analytical expression for the inversion of Generalized Inertia Matrix (GIM), was obtained with the use of DeNOC. The point mass formulation using DeNOC-P [28] was used to identify the dynamic parameters of the simplified model of serial manipulator in [29].

Dynamic parameter identification has attracted researchers due to the use of the model-based controller, accurate off-line programming, and validation of simulation results. The complete dynamic model of a robotic manipulator is highly coupled and represented using a set of non-linear second-order differential equations. However the EOM's is linear with respect to the dynamic parameters, i.e., masses, inertias, mass center locations, etc., and this property is utilized for the dynamic identification. Note that dynamic parameters for a given rigid link of robot consist of ten standard inertial parameters (SIP), namely, six components of the moment of inertia tensor, three components of the first moment of the center of mass and mass of the link. Also, actuators model, if included, will consist of rotor inertia, friction parameters, etc. The knowledge of torque constant and transmission ratio, as shown in Fig. 4.1, are also required to convert a torque to current and rotor rotation. Dynamic identification of robots deals with the estimation of the above parameters which influence the dynamic behavior. The prior knowledge of kinematic parameters is however essential and its identification was discussed in Sect. 4.1.1. The dynamic parameters are used to estimate the driving torques and forces to move the robot.

Note that the value of the inertial parameters could be measured (in the case of the mass) or calculated (in the case of inertia) using CAD models. But due to the limitation in the measurement accuracy and the fact that CAD drawings are not generally modeled with sufficient precision including small items such as screws, washers, wiring, and transmissions. The derived dynamic parameters value from CAD drawing deviate from the actual values. Further, measurements of masses and inertias of the links require dismantling it. The above approaches are impractical. Moreover,

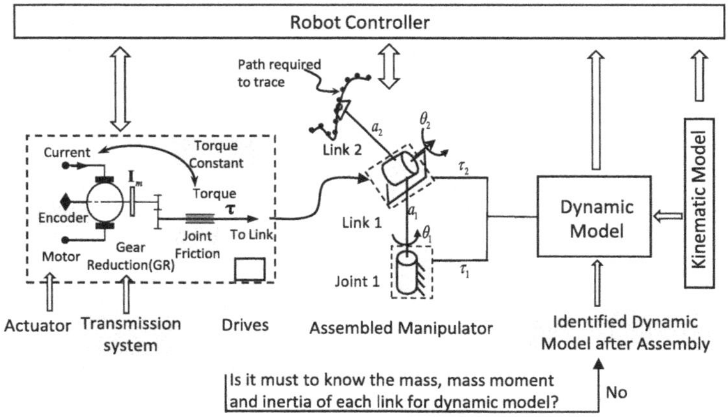

Fig. 4.1 General scheme of a robotic manipulator

the joint friction, etc., cannot be modeled in the CAD environment accurately and henceforth experiments are necessary for the identification of these parameters as well.

In general, a standard robot identification procedure consists of modeling, experiment design and data acquisition, parameter estimation, and model validation [30].

4.1.2.1 Linear Dynamic Model

Among various dynamic modeling techniques, one can select any method. Regardless of the modeling type selected, dynamic parameters appear linearly in EOM. The necessary torque required to drive a robot is written as $\boldsymbol{\tau} = \mathbf{Y}(\boldsymbol{\theta}, \dot{\boldsymbol{\theta}}, \ddot{\boldsymbol{\theta}})\boldsymbol{\chi}_s$, which is linear in SIP $\boldsymbol{\chi}_s$ and the matrix $\mathbf{Y}$ is referred as regressor matrix which is non-linear in joint trajectory parameters, namely, $\boldsymbol{\theta}$, $\dot{\boldsymbol{\theta}}$, and $\ddot{\boldsymbol{\theta}}$. This property facilitates the identification process as the system is linear in unknown parameters. References [31, 32] formulated the robot dynamics using the NE approach which is non-linear with respect to some of the dynamic parameters and transformed into an equivalent model which is linear in the dynamic parameters. Symbolic NE equations were used for the identification of some dynamic parameters for a three link cylindrical robot in [33]. Khalil and Kleinfinger [34] used the EL formulation and regrouped the inertial parameters to get the minimum set of inertial parameters for a tree type systems. The direct calculation of a minimal set of inertial parameters which are sufficient to mimic the dynamics of the robot was expressed for a six-DOF robotic manipulator in [35]. Gibbs–Appell dynamic equations were used to model the dynamics and to ensure physical feasibility of the parameters in [36].

The dynamic parameters can belong to three categories, namely, those which do not affect dynamics, those which affect dynamics in groups and those which affect dynamics individually. The set of minimum number of parameters affecting dynamics is known as base parameters [4]. The dynamic parameter identification of a 3-DOF manipulator was presented in [37]. All the dominant dynamic behavior must be included in the model for accurate prediction of the end-effector (EE) motion. Apart from identification of the SIP, factors such as gear reduction ratio (GRR) in transmission, torque constants, and friction parameters are equally important to model and identify and are discussed below.

4.1.2.2 Kinematic Chain and Torque Constant

The motion is transmitted with the use of actuators and transmission systems (mostly gears, belts, and pulleys) associated to each link. Mostly, each manipulator joint is driven by a single motor and gearbox. In some of the industrial manipulators, there are redundantly driven joints where the motion of one joint is caused by two or more actuators due to mechanical coupling as shown in Fig. 4.2. When the gear ratio is large (typically of the order 120–200 in industrial robots), the motor inertia affects the dynamics. The EOM containing motor inertia terms gets magnified by the square of gear reduction ratio (GRR) [38]. The effect of gear-box was also described in [39]. For the complete system, it is referred to as kinematic chain [40]. Hence, the identification of kinematic chain is necessary.

The torque delivered by the joint motors are just sufficient to enable the robot links to move with the desired trajectory or to be in a static pose. So, it becomes necessary for the controller to calculate current required by the motors to counter balance the torque required for a given trajectory. Generally, the torque required can be calculated from the current by multiplying it with a torque constant factor of a motor. Obtaining the values of torque constant experimentally for a robot is discussed in [41].

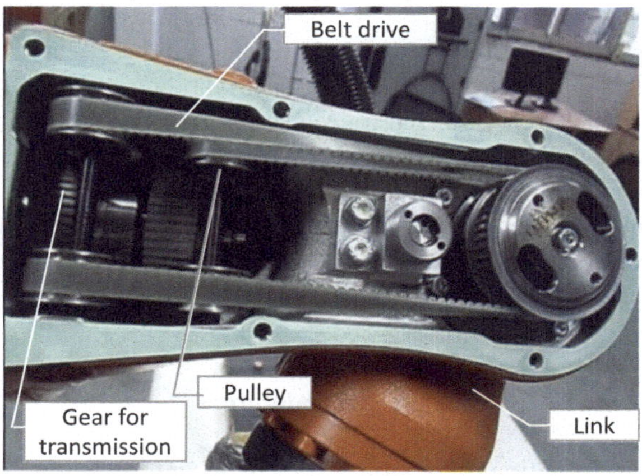

Fig. 4.2 Internal description of KUKA robots showing transmission mechanism of joints 4, 5, and 6 of KUKA KR5 Arc

4.1.2.3 Friction Model

Friction is present between two bodies in relative motion. In robots, it is present in transmission elements, bearings, shaft, etc. The complexity to consider models for individual elements becomes very high. Hence, aggregate friction model for each joint of the robot is considered. In the high GRR robots, the friction phenomenon becomes dominant and accounts for the good percentage of the joint torque. This is not the case in direct drive robots due to the lack of transmission system. Several models are proposed to express the friction characteristics. Torque due to friction for a given joint is usually modeled as a function of its rotational speed, namely, as viscous friction. The non-linear function is mostly described as a sum of dry friction and viscous friction [42, 43]. For achieving the high performance in robots, friction model was used in the EOM in [44]. This friction model was used for identification in [45, 46]. Inclusive dynamic friction model was given for different regimes of friction in [47].

4.1.2.4 Identification Experiment

Suitable trajectory selection is significant and an essential part to improve the accuracy of estimation. It is worth noting that the selection of the exciting trajectory can only account for measured torque uncertainties. Armstrong [48] spotted that the selection of trajectory and the noise in the sensor data significantly affect the identified parameters. Gautier and Khalil [49] used fifth-order polynomial for the excitation trajectory at each joint. The condition number of the regressor matrix in the linear model was used as the criteria to optimize the trajectory. Periodic excitation trajectory based on Fourier series was proposed as excitation trajectory in [50]. Calafiore et al. [51] used a physical constraint of the joint values while optimizing trajectory along with Fisher information matrix to guarantee minimal uncertainty in the parameter estimation. The robust trajectory design approach along with the signal processing of the noisy data to filter out the effect of noise was done in [52]. Note that in most of the installed industrial robots restricts the access to the controller and the desired optimized trajectory for identification is not possible to command. This case will also be dealt where the trajectory is given based on the sudden variation of direction and speed for identifying the dynamic parameters of an industrial manipulator, namely, KUKA KR5 Arc. The condition number criteria suggested in [49] was considered for finally selecting the trajectory and the estimation of the dynamic parameters.

4.1.2.5 Estimation and Validation

The equations of motion written in linear form facilitate to find the unknown dynamic parameters as stated in Sect. 4.1.2.1. With suitably selected excitation trajectory the dynamic parameters are estimated based on the experimental data. It can be performed offline or online as well. In online estimation, the dynamic parameters are estimated and updated along with the experimental data acquired. In time domain, the parameter estimation was shown to be feasible in [53], whereas in the offline method, the data is collected once and the estimation is done offline. The offline method is most widely used in the estimation of robots dynamic parameters, since the mass and inertia properties of the robot do not change frequently with time. Moreover, when estimating the value of the dynamic parameters of the system, different estimation methods can be used. Most widely used is called least squares estimation [54]. To estimate the dynamic parameters here, the non-square matrix obtained by concatenating the experimental data in the linear dynamic equation results in an overdetermined system. Then the minimization of error problem is solved using Moore–Penrose generalized inverse. Many researchers used the least square method for estimating the dynamic parameter.

4.1.3 Objectives

Based on the background studies the objective of the identification is listed below:

- Identify the kinematic parameters of a serial robot to be used in the precision-based assembly task.
- Develop a dynamic model using DeNOC for dynamic identification.
- Implement the identified model for gaining the complaint robot using the control strategies.

4.2 Kinematic Identification

In this section, the formulation to estimate the DH parameters of a robot is derived using Circle Point Method (CPM). The definitions of the DH parameters used here are provided in Appendix B.2. Also, for the inverse kinematics of the robot used in the next section, i.e., KUKA KR5 Arc are detailed in [55]. The axis information needs to be estimated first which is explained next.

4.2.1 Extraction of Joint Axes

Axis of each joint can be typically defined by its direction cosines and a point on it. In order to extract the joint axes information, a method based on singular value decomposition (SVD) is proposed here, which allows one to quantify the quality of measurement as well. For this, SVD of the matrix associated to the measurement of data obtained from the circle points was carried out. Circle points are those lying

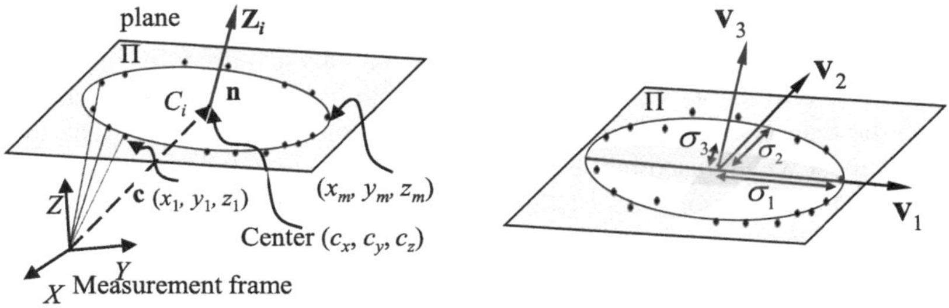

(a) Points w.r.t measurement frame (b) Singular values and normal direction

Fig. 4.3 Joint axis with its center of a circle fitted on a plane

on a trajectory traced by the EE of a robot when only one joint axis is moved as illustrated in Fig. 4.3a. The joint axis information was then obtained. It was used to determine the required DH parameters of the robot under study. An axis of a robot's joint, say, a revolute joint, is typically defined by its direction and a point on it.

Upon actuating a joint, while keeping the rest of the joints locked, the EE will trace a circle. The axis of the circle is the normal direction Z_i to the plane of rotation of the EE. A point on the axis of the circle is the center of the circle C_i. Note that the actual points traced by the EE lie above and below the plane. This is mainly due to the minor inaccuracies in the robot and measurement errors. The mathematical analysis of the measured data provides the joint axis information.

The three-dimensional (3D) coordinates (x_i, y_i, z_i) of the points traced by the EE were stacked in a matrix **K**, as given in Eq. (4.1). The mean of the measured data points $\bar{\mathbf{x}}$ was subtracted from the elements of **K** to get matrix **N**. It contains the transformed data set with mean as zero. Geometrically, all the data points got shifted with their mean at the origin of the measurement frame. Matrices **K** and **N** are shown below:

$$\mathbf{K} \equiv \begin{bmatrix} x_1 & y_1 & z_1 \\ \vdots & \vdots & \vdots \\ x_m & y_m & z_m \end{bmatrix}_{m\times 3}; \ \mathbf{N} \equiv \begin{bmatrix} x_1 - \bar{x} & y_1 - \bar{y} & z_1 - \bar{z} \\ \vdots & \vdots & \vdots \\ x_m - \bar{x} & y_m - \bar{y} & z_m - \bar{z} \end{bmatrix}_{m\times 3} \tag{4.1}$$

where

$$\bar{\mathbf{x}} \equiv \left[\bar{x}\ \bar{y}\ \bar{z}\right]^T = \frac{1}{m}\left[\sum x_i\ \sum y_i\ \sum z_i\right]^T$$

and m being the number of measurements. Applying SVD [56] on matrix **N** one gets two orthogonal matrices **U** and **V** with $m \times 3$ matrix **D** as

$$\mathbf{N} = \mathbf{U}_{m\times m}\mathbf{D}_{m\times 3}\mathbf{V}_{3\times 3} \tag{4.2}$$

Note that the three non-zero diagonal elements of **D**, namely, $\sigma_1,\ \sigma_2$ and σ_3 are the singular values (SV) arranged in decreasing order, i.e., $\sigma_1 > \sigma_2 > \sigma_3$. Geometrical interpretation of these values are shown in Fig. 4.3b, whereas the ratio of the largest to smallest singular values, i.e., σ_1/σ_3 indicates the quality of the acquired data.

For the matrix $\mathbf{V} \equiv [\mathbf{v}_1\ \mathbf{v}_2\ \mathbf{v}_3]$, $\mathbf{v}_3$ denotes the joint axis direction represented by unit vector $\mathbf{n}$ in Fig. 4.3a. The first two columns of $\mathbf{V}$, i.e., $\mathbf{v}_1$ and $\mathbf{v}_2$, form the orthonormal basis for the plane Π. Projecting the 3D coordinates of points on to the plane Π,

$$\mathbf{R}_{m\times 3} = \mathbf{N}_{m\times 3}\mathbf{V}_{3\times 3} \tag{4.3}$$

The first two columns of $\mathbf{R} \equiv [\mathbf{x}'\ \mathbf{y}'\ \mathbf{z}']$, i.e., $\mathbf{x}'$ and $\mathbf{y}'$, contain the x and y-coordinates of the points on the plane Π that are indicated in Fig. 4.3b. Geometrically, they should be on the circle whose center and radius can be obtained using the following equation of the circle:

$$\begin{aligned} &(x-c_1)^2 + (y-c_2)^2 = r^2 \\ &\text{or,}\ \ 2xc_1 + 2yc_2 + k_3 = x^2 + y^2 \\ &\text{where,}\ \ k_3 \equiv r^2 - c_1^2 - c_2^2 \end{aligned} \tag{4.4}$$

Substituting the values of x- and y-coordinates from $\mathbf{x}'$ and $\mathbf{y}'$ of matrix $\mathbf{R}$, three necessary parameters of the circle, namely, c_1, c_2 and r, can be obtained as the solution to the following set of linear algebraic equations:

$$\underbrace{\begin{bmatrix} 2x'_1 & 2y'_1 & 1 \\ \vdots & \vdots & \vdots \\ 2x'_m & 2y'_m & 1 \end{bmatrix}}_{\mathbf{A}_{m\times 3}} \underbrace{\begin{bmatrix} c_1 \\ c_2 \\ k_3 \end{bmatrix}}_{\mathbf{y}_{3\times 1}} = \underbrace{\begin{bmatrix} x'^2_1 + y'^2_1 \\ \vdots \\ x'^2_m + y'^2_m \end{bmatrix}}_{\mathbf{b}_{3\times 1}} \tag{4.5}$$

Using the Moore–Penrose generalized inverse of matrix $\mathbf{A}$ [56], the least square solution to the unknown vector $\mathbf{y}$ is obtained as $\mathbf{y} = \mathbf{A}^{\dagger}\mathbf{b}$, where $\mathbf{A}^{\dagger} \equiv (\mathbf{A}^T\mathbf{A})^{-1}\mathbf{A}^T$. Note that the coordinates of the center of the circle, i.e., (c_1, c_2), is on the plane Π spanned by $[\mathbf{v}_1\ \mathbf{v}_2]$. The coordinates of the center $\mathbf{c} \equiv [c_x, c_y, c_z]^T$ were identified on the basis of the calculated values of c_1, c_2 and $\mathbf{v}_1, \mathbf{v}_2$ along with the addition of the mean vector $\bar{\mathbf{x}}$, i.e.,

$$\mathbf{c} = c_1\mathbf{v}_1 + c_2\mathbf{v}_2 + \bar{\mathbf{x}} \tag{4.6}$$

The joint axis direction $\mathbf{n}$ and the center $\mathbf{c}$ are the inputs for the extraction of DH parameters.

4.2.2 *Extraction of DH Parameters*

The information of all the joint axes was obtained using the methodology explained in the previous section. Next, the assignment of the base frame based on the rotation of joint 1 and the points on the robot's base was done. This is shown in Fig. 4.4a.

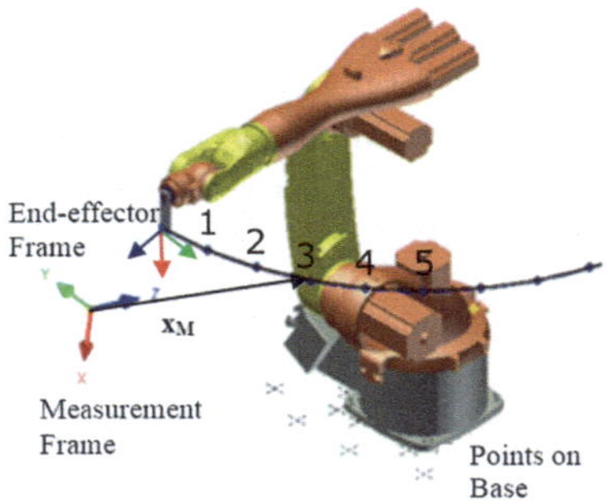

(a) Points w.r.t. measurement frame

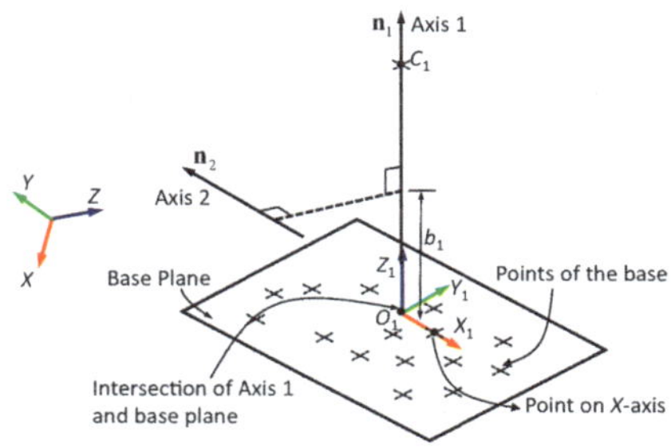

(b) Base frame assignment

Fig. 4.4 Center and a joint axis direction

The rotation of joint 1 gives the center C_1 and normal $\mathbf{n}_1$. Figure 4.4b shows the points on the base were used to fit a plane ($ax + by + cz + d = 0$). The intersection of the 1st joint axis, denoted as $\mathbf{z}_1$ and the plane was marked as the origin O_1. This was followed by the assignment of X_1-axis by joining origin O_1 to any point on the base plane. The axis Y_1 was then defined as a right-handed triad. For the extraction of the DH parameters, note that the center $\mathbf{c} \equiv [c_x, c_y, c_z]^T$ and the axis of rotation direction $\mathbf{n} \equiv [n_x, n_y, n_z]^T$ for each joint were known from the SVD of the measured data. Using the approach presented in [57], vectors $\mathbf{c}$ and $\mathbf{n}$ can then be combined to represent a dual vector in the form, $\hat{\mathbf{n}}_i = \mathbf{n}_i + \varepsilon(\mathbf{c}_i \times \mathbf{n}_i)$, where $\varepsilon^2 = 0$. The formulation of extracting of DH parameters using dual vectors is explained in Appendix B.3.

4.3 Dynamic Identification

Figure 4.5 depicts the scheme of dynamics identification which involves modeling, experiment design, data acquisition with signal processing and parameter estimation with validation. Here, two approaches to estimate the dynamic parameters of a robot are illustrated. First, in a simulation environment using the computer aided design technique that uses the CAD model of the robot in `RoboAnalyzer` [58]. Presently, the coupled mechanism of the wrist joint is not accounted in the RoboAnalyzer software. The inverse dynamics solver uses the DeNOC formulation based on Newton–Euler equations of motion and an orthogonal complement matrix algorithm [59] which is implemented in the inverse dynamics (idyn) module of the software. The inverse dynamics module is utilized for getting the torque and joint angles data required as input in dynamic identification. This is useful for validating the linear in parameters (LIP) dynamic identification model. The second one is using experimental readings of KUKA KR5 Arc installed in the PAR Lab, IIT Delhi.

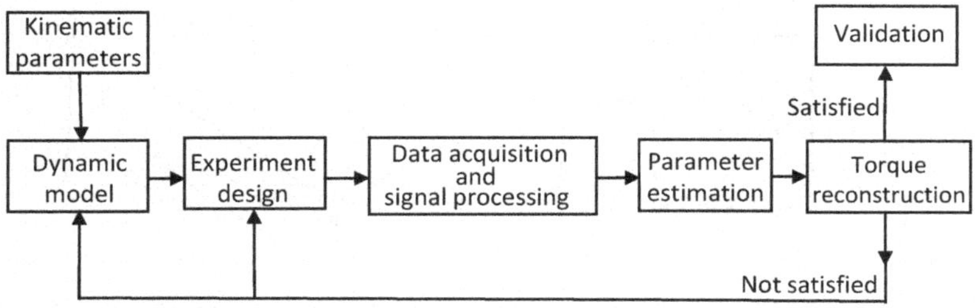

Fig. 4.5 Steps in dynamic parameter identification

4.3.1 Dynamic Modeling

The dynamic modeling based on Decoupled Natural Orthogonal Matrices (DeNOC) begins with the unconstrained Newton-Euler (NE) equation of motion (EOM). This section outlines the steps for a general n-link serial chain robot shown in Fig. 4.6a. Firstly, the NE equations were formulated for a serial kinematic chain. Next, it was suitably modified to identify the unknown dynamic parameters.

4.3.2 Newton–Euler Equations of Motion

Euler's and Newton's equations of motion of the ith rigid link in a serial chain of Fig. 4.6a are written from its free body diagram shown in Fig 4.6b as

$$\mathbf{I}_i^c \dot{\boldsymbol{\omega}}_i + \tilde{\boldsymbol{\omega}}_i \mathbf{I}_i^c \boldsymbol{\omega}_i = \mathbf{n}_i^c \tag{4.7}$$

$$m_i \ddot{\mathbf{c}}_i = \mathbf{f}_i^c \tag{4.8}$$

where $\mathbf{n}_i^c$ is the resultant external moments and $\mathbf{f}_i^c$ is the resultant force acting about the link's center of mass (COM), C_i. Moreover, $\mathbf{I}_i^c$ is the inertia tensor with respect to C_i. The angular velocity of the link is defined by $\boldsymbol{\omega}_i \equiv [\omega_x, \omega_y, \omega_z]_i^T$ and its linear acceleration is $\ddot{\mathbf{c}}_i \equiv [\ddot{c}_x, \ddot{c}_y, \ddot{c}_z]_i^T$. Furthermore, the 3×3 cross-product matrix associated with the angular velocity $\boldsymbol{\omega}_i$, i.e., $\tilde{\boldsymbol{\omega}}_i$ is introduced as

$$\tilde{\boldsymbol{\omega}}_i \equiv \boldsymbol{\omega}_i \times \mathbf{1} \equiv \begin{bmatrix} 0 & -\omega_z & \omega_y \\ \omega_z & 0 & -\omega_x \\ -\omega_y & \omega_x & 0 \end{bmatrix}_i \tag{4.9}$$

Note that the permissible motion between the two constitutive links is due to the joint whose axis is passing through link origin O_i as shown in Fig. 4.6b along the unit vector $\mathbf{e}_i$. In order to express the NE equations of the ith link with respect to the origin of the link, O_i, the velocity, $\dot{\mathbf{o}}_i$, and the acceleration, $\ddot{\mathbf{o}}_i$, are to be expressed by taking the derivatives of position vector $\mathbf{c}_i = \mathbf{o}_i + \mathbf{d}_i$ as

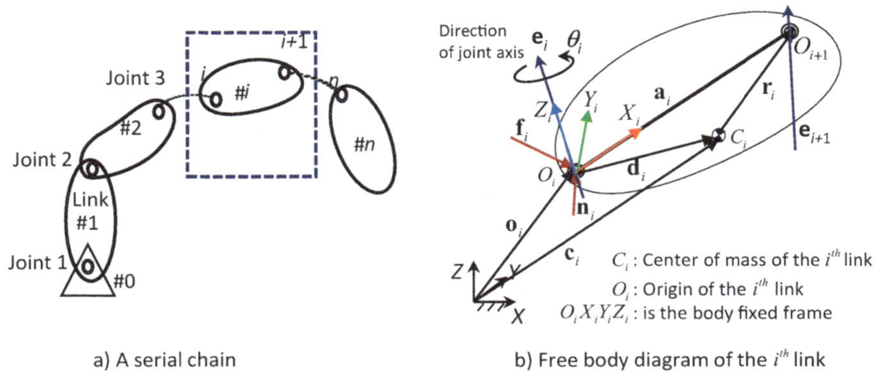

Fig. 4.6 A serial-chain robot with free body diagram of ith link

$$
\begin{aligned}
\dot{\mathbf{c}}_i &= \dot{\mathbf{o}}_i + \boldsymbol{\omega}_i \times \mathbf{d}_i \\
\ddot{\mathbf{c}}_i &= \ddot{\mathbf{o}}_i + \dot{\boldsymbol{\omega}}_i \times \mathbf{d}_i + \boldsymbol{\omega}_i \times (\boldsymbol{\omega}_i \times \mathbf{d}_i)
\end{aligned} \tag{4.10}
$$

Also, the inertia tensor about O_i, namely, $\mathbf{I}_i$, resultant force, $\mathbf{f}_i$, and the resultant moment $\mathbf{n}_i$ about O_i are as follows:

$$
\begin{aligned}
\mathbf{I}_i^c &= \mathbf{I}_i + m_i \tilde{\mathbf{d}}_i \tilde{\mathbf{d}}_i \\
\mathbf{f}_i^c &= \mathbf{f}_i \\
\mathbf{n}_i^c &= \mathbf{n}_i - \tilde{\mathbf{d}}_i \mathbf{f}_i
\end{aligned} \tag{4.11}
$$

where $\tilde{\mathbf{d}}_i$ is the 3×3 skew-symmetric cross-product matrix associated with the vector $\mathbf{d}_i$. Substituting Eqs. (4.10) and (4.11) into Eqs. (4.7) and (4.8), the NE equations of motion for the ith link can be represented with respect to the origin O_i as

$$\mathbf{n_i} = \mathbf{I}_i \dot{\boldsymbol{\omega}}_i + m_i \mathbf{d}_i \times \ddot{\mathbf{o}}_i + \boldsymbol{\omega}_i \times \mathbf{I}_i \boldsymbol{\omega}_i \tag{4.12}$$

$$\mathbf{f_i} = m_i \ddot{\mathbf{o}}_i - m_i \mathbf{d}_i \times \dot{\boldsymbol{\omega}}_i - \boldsymbol{\omega}_i \times (m_i \mathbf{d}_i \times \boldsymbol{\omega}_i) \tag{4.13}$$

The above equations of motion can then be written in compact form in terms of the six-dimensional (6D) twist $\mathbf{t}_i \equiv [\boldsymbol{\omega}_i^T \; \mathbf{v}_i^T]^T$, twist-rate $\dot{\mathbf{t}}_i = [\dot{\boldsymbol{\omega}}_i^T \; \dot{\mathbf{v}}_i^T]^T$ and wrench $\mathbf{w} = [\mathbf{n}_i^T \; \mathbf{f}_i^T]^T$ [60] as

$$\mathbf{M}_i \dot{\mathbf{t}}_i + \mathbf{W}_i \mathbf{M}_i \mathbf{E}_i \mathbf{t}_i = \mathbf{w}_i \tag{4.14}$$

where $\mathbf{M}_i$, $\mathbf{W}_i$ and $\mathbf{E}_i$ are the 6×6 matrices of mass, angular velocities, and elementary matrix, respectively, as

$$\mathbf{M}_i \equiv \begin{bmatrix} \mathbf{I}_i & m_i \tilde{\mathbf{d}}_i \\ -m_i \tilde{\mathbf{d}}_i & m_i \mathbf{1} \end{bmatrix} \quad \mathbf{W}_i \equiv \begin{bmatrix} \tilde{\boldsymbol{\omega}}_i & \mathbf{O} \\ \mathbf{O} & \tilde{\boldsymbol{\omega}}_i \end{bmatrix} \text{ and } \mathbf{E}_i \equiv \begin{bmatrix} \mathbf{1} & \mathbf{O} \\ \mathbf{O} & \mathbf{O} \end{bmatrix} \tag{4.15}$$

For estimating the unknown dynamic parameters, the equations of motion must be rearranged by isolating the unknown dynamic parameters, which is presented next.

4.3.3 Linear Dynamic Model

Note in Eq. (4.14), that the inertia tensor $\mathbf{I}_i$ is configuration-dependent and each element of it varies with the orientation of the link if represented in the base frame. However, the inertia tensor expressed in the link-fixed frame i has an invariant element. Thus, it is not required to recalculate the inertia tensor when link #i gets reoriented. It is preferred to write inertia of the ith-link about O_i using the dyadic form as explained in Appendix B.6. Equations (4.12) and (4.13) are rewritten as

$$\mathbf{n}_i = \boldsymbol{\Omega}(\dot{\boldsymbol{\omega}}_i)\bar{\mathbf{I}}_i - \ddot{\mathbf{o}}_i\bar{\mathbf{d}}_i + \tilde{\boldsymbol{\omega}}_i\boldsymbol{\Omega}(\boldsymbol{\omega}_i)\bar{\mathbf{I}}_i \quad (4.16)$$

$$\mathbf{f}_\mathrm{i} = (\ddot{\mathbf{o}}_i)m_i + \dot{\tilde{\boldsymbol{\omega}}}_i\bar{\mathbf{d}}_i + \tilde{\boldsymbol{\omega}}_i^2\bar{\mathbf{d}}_i \quad (4.17)$$

where $\boldsymbol{\Omega}(\dot{\boldsymbol{\omega}}_i) \equiv \left[\mathbf{ii}^T\dot{\boldsymbol{\omega}}_i\ (\mathbf{ij}^T+\mathbf{ji}^T)\dot{\boldsymbol{\omega}}_i\ (\mathbf{ik}^T+\mathbf{ki}^T)\dot{\boldsymbol{\omega}}_i\ \mathbf{jj}^T\dot{\boldsymbol{\omega}}_i\ (\mathbf{jk}^T+\mathbf{kj}^T)\dot{\boldsymbol{\omega}}_i\ \mathbf{kk}^T\dot{\boldsymbol{\omega}}_i\right]$ (refer Appendix B.6). The unknown dynamic parameters from Eq. 4.14 were isolated in Eqs. 4.16 and 4.17. The 6 independent inertial element were stacked in 6D vector $\bar{\mathbf{I}} \equiv [I_{xx}\ I_{xy}\ I_{xz}\ I_{yy}\ I_{yz}\ I_{zz}]^T$, the three mass moment component as 3×1 vector, $\bar{\mathbf{d}}_i \equiv [md_x\ md_y\ md_z]_i^T$ and the mass m_i. Hence, altogether 10 unknown dynamic parameters can be stacked in the 10-dimensional vector $\boldsymbol{\chi}_i$ as

$$\boldsymbol{\chi}_i \equiv \left[I_{xx}, I_{xy}, I_{xz}, I_{yy}, I_{yz}, I_{zz}, md_x, md_y, md_z, m\right]_i^T \quad (4.18)$$

Equations (4.16) and (4.17) are combined to form the 6D wrench $\mathbf{w}_i$ as

$$\mathbf{w}_i = \mathbf{V}_i\boldsymbol{\chi}_i \quad (4.19)$$

$$\mathbf{V}_i = \begin{bmatrix} (\boldsymbol{\Omega}(\dot{\boldsymbol{\omega}}_i) + \tilde{\boldsymbol{\omega}}_i\boldsymbol{\Omega}(\boldsymbol{\omega}_i))_{3\times6} & \ddot{\mathbf{o}}_{3\times3} & \mathbf{0}_{3\times1} \\ \mathbf{0}_{3\times6} & (\dot{\tilde{\boldsymbol{\omega}}}_i + \tilde{\boldsymbol{\omega}}_i^2)_{3\times3} & \ddot{\mathbf{o}}_{3\times1} \end{bmatrix} \quad (4.20)$$

In Eq. 4.20, $\mathbf{V}_i$ is the 6×10 matrix, whereas $\mathbf{w}_i \equiv \mathbf{w}_i^D + \mathbf{w}_i^E + \mathbf{w}_i^C = \mathbf{V}_i\boldsymbol{\chi}_i$ in which $\mathbf{w}_i^D$, is the wrench due to driving or actuating moments and forces, $\mathbf{w}_i^E$ and $\mathbf{w}_i^C$ are the external and constraint wrenches, respectively. The external wrench, $\mathbf{w}_i^E$, is due to the moments and forces of gravity, friction, joint actuators, environmental effects, etc. The constraint wrench, $\mathbf{w}_i^C$, is due to the presence of reaction moments and forces at the joint interfaces. Writing Eq. (4.20) for all the n connected bodies as

$$\mathbf{w} = \mathbf{w}^D + \mathbf{w}^E + \mathbf{w}^C = \mathbf{V}\boldsymbol{\chi} \quad (4.21)$$

where the matrix $\mathbf{V}$ is the block diagonal, $\mathbf{V}_i$ and the vectors $\mathbf{w}^D$, $\mathbf{w}^E$, $\mathbf{w}^C$ and $\boldsymbol{\chi}$ are obtained by stacking the corresponding vectors for all bodies.

Before solving Eq. (4.21) for estimating the unknown dynamic parameters, it is essential to modify the equations of motion according to the external and constraint wrench acting on the system. They are discussed next.

4.3.4 *Constraint Wrench*

The power due to constraint wrenches is equal to zero [61]. This means that the vector of constraint moments and forces are orthogonal to the columns of the velocity transformation matrix. For the sake of completeness, the velocity transformation, i.e., DeNOC matrices are derived using the kinematic constraints as shown below.

4.3.4.1 Constraint Kinematic Equation

Considering $\boldsymbol{\omega}_i$ and $\mathbf{v}_i$ as the angular velocity and the linear velocity of the origin point, O_i, of the ith link, respectively, as shown in Fig. 4.6b. The twist, $\mathbf{t} \equiv [\boldsymbol{\omega}^T \, v^T]^T$, is expressed in terms of the twist of its previous body, i.e., the $(i-1)$st one, as in [60], i.e.,

$$\mathbf{t}_i = \mathbf{A}_{i,i-1}\mathbf{t}_{i-1} + \mathbf{p}_i\dot{\theta}_i \tag{4.22}$$

where $\mathbf{A}_{i,i-1}$ is the 6×6 twist propagation matrix which transforms the twist of the $(i-1)^{st}$ body to the ith one. The (6×6) matrix $\mathbf{A}_{i,i-1}$, and the 6D vector $\mathbf{p}_i$, are referred as the twist propagation matrix and the joint motion propagation vector, respectively. They are given by

$$\mathbf{A}_{i,i-1} \equiv \begin{bmatrix} \mathbf{1} & \mathbf{O} \\ \tilde{\mathbf{a}}_{i,i-1} & \mathbf{1} \end{bmatrix}; \mathbf{p}_i \equiv \begin{bmatrix} \mathbf{e}_i \\ \mathbf{0} \end{bmatrix} \text{ (revolute)} \tag{4.23}$$

where $\tilde{\mathbf{a}}_{i,i-1}$ is the skew symmetric matrix associated with the vector $\mathbf{a}_{i,i-1} \equiv [a_{i,i-1x}, a_{i,i-1y}, a_{i,i-1z}]^T$, and $\mathbf{e}_i$ is the unit vector along the axis of rotation or translation of the ith joint. The notations **1**, **O**, and **0** in Eq. (4.23) represent identity matrix, null matrix, and null vector whose sizes are compatible to the expressions where they appear. The detailed physical interpretation of these matrices are given in [59]. The generalized twist and the joint rates for the n-link serial chain robot denoted with twist **t** is given as

$$\mathbf{t} \equiv \begin{bmatrix} \mathbf{t}_1^T & \cdots & \mathbf{t}_i^T & \cdots & \mathbf{t}_n^T \end{bmatrix}^T, \text{ and } \dot{\boldsymbol{\theta}} \equiv \begin{bmatrix} \dot{\theta}_1 & \cdots & \dot{\theta}_i & \cdots & \dot{\theta}_n \end{bmatrix}^T \tag{4.24}$$

where **t** and θ are $6n$ and n-dimensional vectors obtained after stacking twist and joint angles of all the links. Then, the expression for the generalized twist can be obtained from Eq. (4.23) and Eq. (4.24) as

$$\mathbf{t} = \mathbf{A}\mathbf{t} + \mathbf{N}_d\dot{\boldsymbol{\theta}} \tag{4.25}$$

The twist can easily be written in terms of the joint rates as

$$\mathbf{t} = (\mathbf{1} - \mathbf{A})^{-1}\mathbf{N}_d\dot{\boldsymbol{\theta}} \equiv \mathbf{N}_l\mathbf{N}_d\dot{\boldsymbol{\theta}} = \mathbf{N}\dot{\boldsymbol{\theta}} \tag{4.26}$$

The matrices $\mathbf{N}_l$ and $\mathbf{N}_d$ are referred as Decoupled Natural Orthogonal Complement (DeNOC) matrices reported in [62]. These are the decoupled form of the $6n \times n$ Natural Orthogonal Complement (NOC) matrix $\mathbf{N}$ [61], which together are referred to as the DeNOC matrices. These are given by

$$\mathbf{N}_l \equiv \begin{bmatrix} \mathbf{1} & \mathbf{O} & \cdots & \mathbf{O} \\ \mathbf{A}_{21} & \mathbf{1} & \cdots & \mathbf{O} \\ \vdots & \vdots & \ddots & \vdots \\ \mathbf{A}_{n1} & \mathbf{A}_{n2} & \cdots & \mathbf{1} \end{bmatrix} \text{ and } \mathbf{N}_d \equiv \begin{bmatrix} \mathbf{p}_1 & \mathbf{0} & \cdots & \mathbf{0} \\ \mathbf{0} & \mathbf{p}_2 & \cdots & \mathbf{0} \\ \vdots & \vdots & \ddots & \vdots \\ \mathbf{0} & \mathbf{0} & \cdots & \mathbf{p}_n \end{bmatrix} \tag{4.27}$$

where $\mathbf{N}_l$ is the $6n \times 6n$ matrix and $\mathbf{N}_d$ is the $6n \times n$ matrix. The columns of the velocity constraint matrix, i.e., the NOC or the resulting DeNOC matrices are orthogonal to the vectors of the constraint moments and forces. Hence, the projection of the constraint joint wrench onto the joint axes by pre-multiplication of Eq. (4.21) gives the minimal set of equations of motion. Specifically, the result, $\mathbf{N}_d^T\mathbf{N}_l^T\mathbf{w}^C = \mathbf{0}$ will yield:

$$\mathbf{N}_d^T\mathbf{N}_l^T(\mathbf{w}^D + \mathbf{w}^E + \mathbf{w}^C) = \mathbf{N}_d^T\mathbf{N}_l^T\mathbf{V}\boldsymbol{\chi} \tag{4.28}$$

$$\boldsymbol{\tau} + \boldsymbol{\tau}^E = \mathbf{Y}\boldsymbol{\chi} \tag{4.29}$$

where $\boldsymbol{\tau}$ is the vector of driving joint torques, $\mathbf{Y}$ is the upper triangular matrix and $\boldsymbol{\chi}$ is the 10-dimensional vector termed as standard inertial parameters (SIP) of each link stacked one below other as:

$$\mathbf{Y} = \begin{bmatrix} \mathbf{p}_1^T\mathbf{V}_1 & \mathbf{p}_2^T\mathbf{A}_{12}^T\mathbf{V}_2 & \cdots & \mathbf{p}_n^T\mathbf{A}_{1n}^T\mathbf{V}_n \\ \mathbf{0}^T & \mathbf{p}_2^T\mathbf{V}_2 & \cdots & \mathbf{p}_n^T\mathbf{A}_{2n}^T\mathbf{V}_n \\ \vdots & \vdots & \ddots & \vdots \\ \mathbf{0}^T & \mathbf{0}^T & \mathbf{0}^T & \mathbf{p}_n^T\mathbf{V}_n \end{bmatrix} \equiv \begin{bmatrix} \mathbf{y}_{11}^T & \mathbf{y}_{12}^T & \cdots & \mathbf{y}_{1n}^T \\ \mathbf{0}^T & \mathbf{y}_{22}^T & \cdots & \mathbf{y}_{2n}^T \\ \vdots & \vdots & \ddots & \vdots \\ \mathbf{0}^T & \mathbf{0}^T & \cdots & \mathbf{y}_{nn}^T \end{bmatrix} \tag{4.30}$$

In Eq. (4.29), $\boldsymbol{\chi} \equiv [\boldsymbol{\chi}_1^T, \ \boldsymbol{\chi}_2^T, \cdots, \boldsymbol{\chi}_n^T]^T$. Generally, the joint angles and currents or torques data are the measured quantity available for a given trajectory from the manipulator. Hence, eliminating the constraint wrench as in Eq. (4.29) will be helpful in the identification of unknown dynamic parameters using the driving joint torque information only ignoring the other reaction forces which are generally not measured at the joints. Torque of the ith link can also be written in summation form as

$$\tau_i = \sum_{j=i}^{n} \mathbf{y}_{ij}^T \chi_i \tag{4.31}$$

where n is the number of links.

4.3.4.2 External Wrench

The moments and forces due to gravitational acceleration, friction, and torques due to motor inertia are three typical examples of external forces that one needs to take care of, particularly, for the robotic systems analyzed in this work. To include these effects in the torque expression $\boldsymbol{\tau}^E$ of Eq. (4.29), the following conditions are made.

Acceleration due to gravity

The wrench on link i due to gravity is written as

$$\mathbf{w}_i^g = \begin{bmatrix} m_i \tilde{\mathbf{d}}_i \mathbf{g} \\ m_i \mathbf{g} \end{bmatrix} \tag{4.32}$$

The above expression contains mass and mass moment terms. It is incorporated in Eq. (4.19) with the acceleration $\ddot{\mathbf{o}}_i$ of link's origin O_i as

$$\mathbf{w}_i = \underbrace{\begin{bmatrix} \boldsymbol{\Omega}(\dot{\boldsymbol{\omega}}_i) + \tilde{\boldsymbol{\omega}}_i \boldsymbol{\Omega}(\boldsymbol{\omega}_i) & \ddot{\mathbf{o}}_i - \mathbf{g} & \mathbf{0} \\ \mathbf{0}_{3\times 6} & \dot{\tilde{\boldsymbol{\omega}}}_i + \tilde{\boldsymbol{\omega}}_i^{\,2} & \ddot{\mathbf{o}}_i - \mathbf{g} \end{bmatrix}}_{\mathbf{V}_{g,i}} \chi_i \tag{4.33}$$

The above expression are pre-multiplied with the transpose of the DeNOC matrices to obtain Eq. 4.29. Similarly, the equations of motion are modified to incorporating the torque due to friction at each joint and motor inertia.

Torque due to friction and rotor inertia of actuator

Friction force resists the relative motion between the surfaces. It is normally defined as a product of the coefficient of friction and the normal force. Friction in robot with revolute joints are mostly modeled as load independent. Therefore, friction is modeled as a joint torque which is a function of angular joint velocities. Figure 4.7a shows a simple friction model given by Newton as a function of Coulomb and viscous frictions that depend upon the direction and the magnitude of angular velocity, respectively, as [63]

$$\tau_{f,i} = sgn(\dot{\theta}_i) f_{ci} + \dot{\theta}_i f_{vi} \tag{4.34}$$

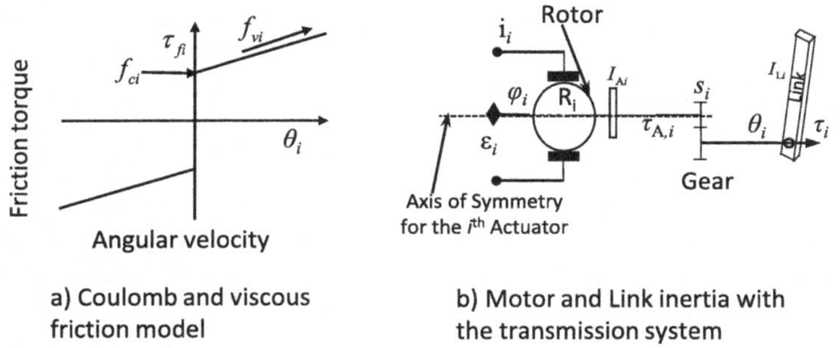

Fig. 4.7 Friction model and the motor with transmission system

where f_{ci} is the coefficient of the Coulomb friction while f_{vi} is the viscous friction and sgn is the signum function which returns the sign of a real number. As friction is nonconservative force at the manipulator joints, it is then subtracted from the actuation torques as $\tau_i - \tau_{fi}$.

The rotor inertia properties can also be included in the dynamic model. Assuming that the rotor has a symmetric mass distribution about its axis of rotation which is generally true with rotary actuators. As a result, the constant scalar moment of inertia of the rotor $I_{\text{A}i}$ about its rotation axis will contribute to the actuation torque as

$$\tau_{\text{A}i} = s_i^2 I_{\text{A}i} \ddot{\theta}_i \tag{4.35}$$

where s_i is the gear ratio of the transmission system and $I_{\text{A}i}$ is the rotor's inertia about its axis of rotation. Note that the motor and friction models considered here are linearly parameterizable. Hence, the unknown parameters f_{ci}, f_{vi} and $I_{\text{A}i}$ are appended in the vector χ which then becomes 13×1 vector as

$$\chi_i \equiv \left[\bar{\mathbf{I}}_i^T \; (m_i \mathbf{d}_i)^T \; m_i \; f_{ci} \; f_{vi} \; I_{\text{A}i} \right]^T \tag{4.36}$$

With the additional terms accounting for friction and rotor inertia that were added in the vector χ of Eq. (4.31), the corresponding expressions to these unknown parameters were also included as $[\,\mathbf{y}_{ij}^T \; sign(\dot{\theta}_j) \; \dot{\theta}_j \; s_j^2 \ddot{\theta}_j\,]$ in the element of the regressor matrix, where j is the number of link. Note that the link #i is now modeled as link with the actuator. Hence, the mass, mass moment, and the inertia terms will also include the actuators contribution to the dynamics. Therefore, $\bar{\mathbf{I}}_i = \bar{\mathbf{I}}_{Li} + \bar{\mathbf{I}}_{\text{A}i}$, $m_i \mathbf{d}_i = m_{Li} \mathbf{d}_{Li} + m_{\text{A}i} \mathbf{d}_{\text{A}i}$ and $m_i = m_{Li} + m_{\text{A}i}$ are, respectively, the overall inertia tensor, mass moment, and mass. Here, subscript L is for link and A is for the actuator. The main effects that were considered in the dynamic modeling were (1) model without the effect of gravity; (2) model including the gravity effect; (3) model with friction and with rotor inertia.

The dynamic model presented in this section incorporating the external wrench will be used in the later section for the experimental identification of unknown parameters. Note that no separate symbols were used to account for the additional effects in Eq. 4.29 which is written as

$$\boldsymbol{\tau} = \mathbf{Y}\boldsymbol{\chi} \equiv \mathbf{Y}(\boldsymbol{\theta}, \dot{\boldsymbol{\theta}}, \ddot{\boldsymbol{\theta}})\boldsymbol{\chi} \tag{4.37}$$

where $\mathbf{Y}(\boldsymbol{\theta}, \dot{\boldsymbol{\theta}}, \ddot{\boldsymbol{\theta}})$ is the regressor matrix of size (number of joints n × number of unknown dynamic parameters), and $\boldsymbol{\chi}$ is a vector of unknown parameters. The number of unknown parameters will be mentioned as per the effects included in the model used, in later sections.

4.3.5 *Identification of Dynamic Parameters*

The dynamic model in Eq. (4.37) is linear in parameters (LIP) form where $\boldsymbol{\chi}$ is nothing but the vector of unknown dynamic parameters. Since the dynamics are written in linear form, linear least squares technique can be used to find out the parameters. However, before doing least squares, investigation and evaluating the non-linear structure of matrix $\mathbf{Y}$ is essential to find the accurate and unique solution. The two factors influencing the identifiability of the dynamic parameters are dealt next.

4.3.5.1 Properties of Regressor Matrix

Investigation of regressor matrix, $\mathbf{Y}$ of Eq. (4.37), involves the evaluation of linearly dependent columns. Numerically, this problem is equivalent to find the rank of the regressor matrix using random values of $\boldsymbol{\theta}$, $\dot{\boldsymbol{\theta}}$, and $\ddot{\boldsymbol{\theta}}$. The singular value decomposition (SVD) of the matrix $\mathbf{Y}$ was used by [64] for this purpose. We have proposed Gauss–Jordan elimination technique for the same. This technique is computationally less expensive than SVD and results in a unique set of base parameters. An example is provided in the Appendix B.7 to illustrate the Gauss–Jordan elimination technique that results into row reduced Echelon form and a new set of unknown parameters. The three situations may arise with this numerical technique. They are as follows:

- A column of the regressor matrix $\mathbf{Y}$ is identically zero for all possible trajectories points. Hence, the unknown dynamic parameter corresponding to the zero column has no effect on the dynamics at all and hence can be removed from the vector of unknown parameters in $\boldsymbol{\chi}$.
- A column of the row reduced form (`rref`) of regressor matrix has only one element in it, i.e., the pivot element, while the other elements in that column are zeros. The unknown dynamic parameter corresponding to this column is independent, and effects, dynamics independently.

- The columns which are linearly dependent on other columns, the former columns can be removed, and a new parameter can replace the parameters of the dependent columns. Hence, these parameters affect the dynamics in a group.

Note that all the dynamic parameters in vector χ cannot be estimated independently, which is equivalent to saying that all of them do not affect the dynamics independently.

4.3.5.2 Excitation Trajectory

Investigation of the evaluation of the excitation trajectory is essential for performing experiments which can result in accurate dynamic parameters in the presence of disturbances such as actuator and measurement noises, etc. The influence of noise on identified parameters was studied using the condition numbers of the regressor matrix, i.e., using `cond`[1] ($\mathbf{Y}$), i.e., effect of the noise is proportional to the condition number. For manipulator with 2-DOF or greater, $cond(\mathbf{Y}) < 100$ indicates that the system gets well excited and as a result gives a good estimate of the dynamic parameters in the presence of measurement noises [4]. An Optimization technique to select the trajectories was reported in [50] where condition number of $\mathbf{Y}$ was minimized for robot identification. However, it is not possible to command the industrial robots according to the calculated optimized trajectory because of other probable constraints, say, obstacles on the robot's way, etc. A simple approach was adopted for this chapter where the persistent excitation by moving the robot using a point to point (`PTP`[2]) motion exciting all the links with sudden change in the angular positions and directions of the EE. The point to movement command is generally available in all the variants of an industrial robot.

The regrouping of the parameters based on Gauss–Jordan elimination method resulted in the minimal set of identifiable parameters. These are known as the base parameters (BPS) represented by χ_b, which allow one to express equation as

$$\tau = \mathbf{Y}_b \chi_b \tag{4.38}$$

where $\mathbf{Y}_b$ is the set of independent columns full rank matrix with dimension $n \times n_b$ where, (n) is the number of joints and (n_b) is the number of BPs .

An excitation trajectory was used to provide joint angles $\boldsymbol{\theta}$ of the joints, the joint velocities $\dot{\boldsymbol{\theta}}$ and the accelerations $\ddot{\boldsymbol{\theta}}$ that must excite the robot under study. For m numbers of joint trajectories and corresponding motor torques, the data were concatenated in Eq. (4.38) as

[1] Command in `MATLAB`®.

[2] Command in Kuka Robotics Language `KRL`.

$$\begin{pmatrix} \tau(t_1) \\ \vdots \\ \tau(t_m) \end{pmatrix} = \begin{pmatrix} \mathbf{Y}_b(\theta(t_1), \dot{\theta}(t_1), \ddot{\theta}(t_1)) \\ \vdots \\ \mathbf{Y}_b(\theta(t_m), \dot{\theta}(t_m), \ddot{\theta}(t_m)) \end{pmatrix} \boldsymbol{\chi}_b \tag{4.39}$$

Interchanging the right and left sides, above equation is rewritten as

$$\mathbf{Y}_{b,\mathrm{T}} \boldsymbol{\chi}_b = \boldsymbol{\tau}_\mathrm{T} \tag{4.40}$$

where the size of the observation matrix with inputs, i.e., $\mathbf{Y}_{b,\mathrm{T}}$, depends on the number of data samples collected from a trajectory. It is $(nm) \times n_b$. Then the BPs can be found using Moore–Penrose generalized pseudoinverse [65] of $\mathbf{Y}_{b,\mathrm{T}}$ as

$$\boldsymbol{\chi}_b = \mathbf{Y}_{b,\mathrm{T}}^{\dagger} \boldsymbol{\tau}_\mathrm{T} \;\; \text{where } \mathbf{Y}_{b,\mathrm{T}}^{\dagger} \equiv (\mathbf{Y}_{b,\mathrm{T}}^T \mathbf{Y}_{b,\mathrm{T}})^{-1} \mathbf{Y}_{b,\mathrm{T}}^T \tag{4.41}$$

Note that the above models only those parameters which affect the dynamics instead of finding all the individual set of inertial parameters for each link. With the reduced number of dynamic parameters in the model, it can be termed as *reduced inverse dynamics identification model.*

4.4 Experiments

In this section, the kinematic and dynamic parameters identification techniques carried out on KUKA KR5 Arc are discussed.

4.4.1 Kinematic Identification of 6-DOF Serial Manipulator

In this section, the DH parameters of the 6-DOF manipulator KUKA KR5 Arc are identified using two approaches. These approaches were classified on the basis of the outcome of simulation results. The first approach is identifying and evaluating the DH parameters found by actuating each link of the robot by restricting its motion within 20°. This was carried out using the state of the art method for measurements, namely, Total Station (TS) and LT. The second approach is by increasing the sector angle with a monocular camera mounted on the robot EE.

The state-of-art measurement devices, i.e., a TS and a LT were used for identification of the DH parameters are discussed. Figure 4.8a shows the TS SOKKIA®-SET230RK. The measurement of the target point on the robot marked as reflector in Fig. 4.8a. It needs to be in the line of sight between reflector point on the robot's EE and the TS. This limits the motion of the joint. Figure 4.8b shows FARO® LT with the

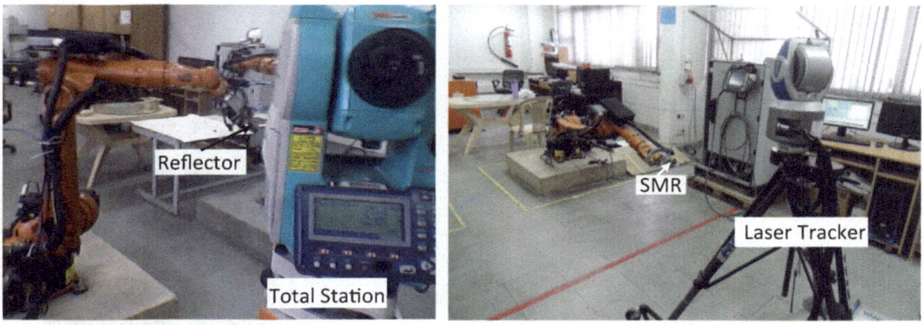

(a) SOKKIA total station (b) FARO laser tracker

Fig. 4.8 Experimental setup for kinematic identification

spherically mounted reflectors (SMR) at the robot's EE. Here also the measurement can be made till the SMR is in the line of sight of the LT which again restricts the joint motions.

4.4.2 Kinematic Identification Using a Monocular Camera

The monocular camera-based technique is adopted here for the pose detection of the robot. Figure 4.10a shows the transformation between the grid, camera, and robot. The trace of the camera by actuating a single joint is also depicted in the Fig. 4.10a. Figure 4.10a shows the fiducial markers used for pose measurement. Note that the transformation matrix $\mathbf{T} = [\mathbf{Q}_{ee}|\mathbf{p}_{ee}]$ maps the camera coordinates to grid/marker coordinates (Fig. 4.10a) where $\mathbf{Q}_{ee}$ is the 3×3 rotation matrix corresponding to the EE's orientation, and $\mathbf{p}_{ee}$ is the position of the EE with respect to the coordinate system defined on the board of markers (Fig. 4.9). One can also obtain the orientation angles roll (A), pitch (B), and yaw (C) from the rotation matrix $\mathbf{Q}_{ee}$ using relations given in [59]. In this way, the pose of the camera mounted on the robot EE w.r.t., the marker frame can be estimated for different poses of the robot.

A Basler pilot camera of resolution (2448×2050) pixels with 8 mm lens was mounted on the EE of an industrial robot KUKA KR5 Arc, to take images of the grid or pattern. The image coordinates were extracted using the methodology proposed in [66] and then supplied to the camera calibration algorithm [67]. An artificial reality toolkit [68] was used to produce boards consisting of array of markers. One of the board that was generated using the library `ARUCO_MIP_36h12` is shown in Fig. 4.9b. Here, all the markers are on the same plane and within a rectangular area.

Another set of experiments were conducted for larger range of joint actuation. The sector angle for the MC setup was increased to 50°. It was done by splitting the set of markers that were pasted at different places such that all of them are not necessarily on the same plane and not within the a fixed rectangular area (Fig. 4.10a). This is

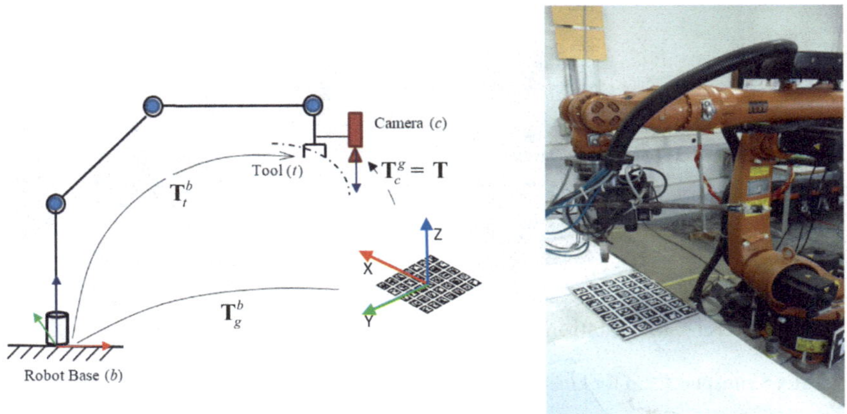

(a) Robot configuration and transformations

(b) Board of fiducial markers created using ArUco and marker coordinate system

Fig. 4.9 Coordinate system for grid and the transformations between the frames

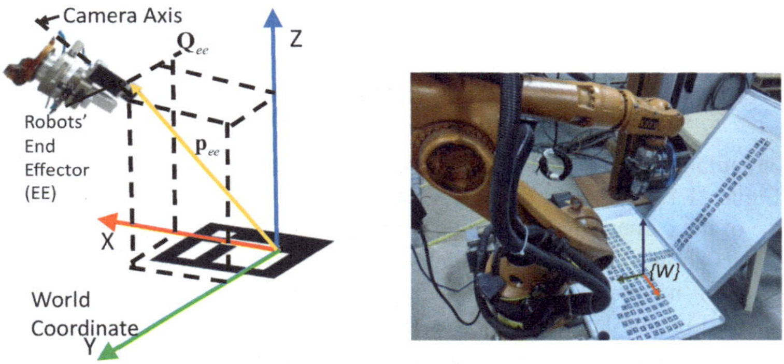

(a) Marker Board

(b) Aruco Marker Map (AMM)

Fig. 4.10 Experimental setup for kinematic identification

called Aruco Marker Maps (AMM) [69]. Using the input images of the markers or marker maps, and the camera calibration parameters, the poses of the camera were obtained (ArUco,OpenCV, [70]).

The increase in actuation angle using AMM gives an advantage of capturing the images in different planes as well. Hence an additional constraints as shown in Fig. 4.11 were used to arrive at the improved values. Since the measurements by the camera would be more accurate if it moves in parallel plane, the following scheme was followed while actuating the joints of the robot.

The architecture of the robot is utilized effectively to refine the length parameters of DH, i.e, b and a. Generally, serial robots are built such that its joints are parallel or

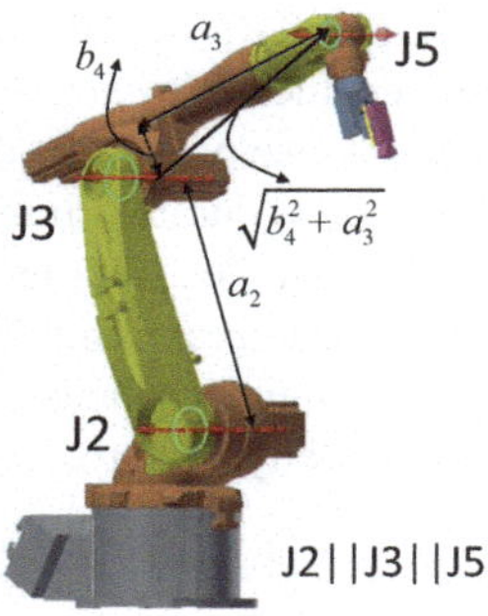

(a) Actuation of joint 2, 3, and 5

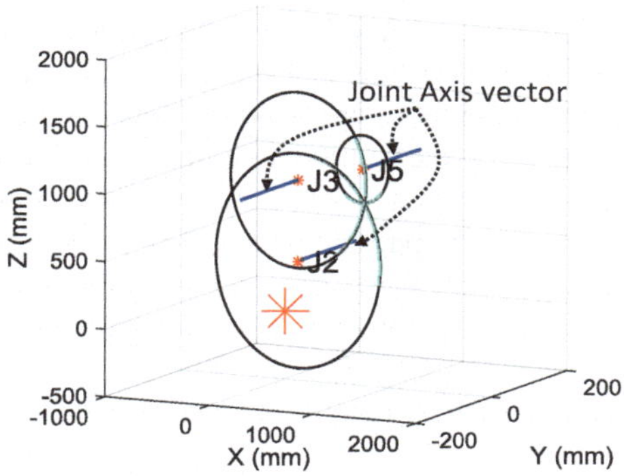

(b) Circles due to rotation of joints 2, 3 and 5

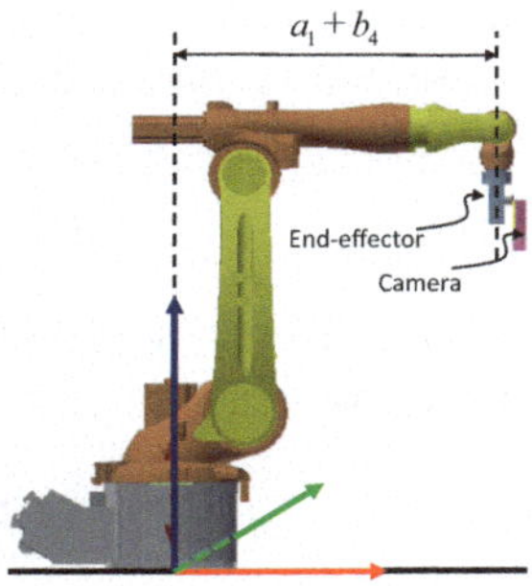

(c) Actuation of joint 1 and 6

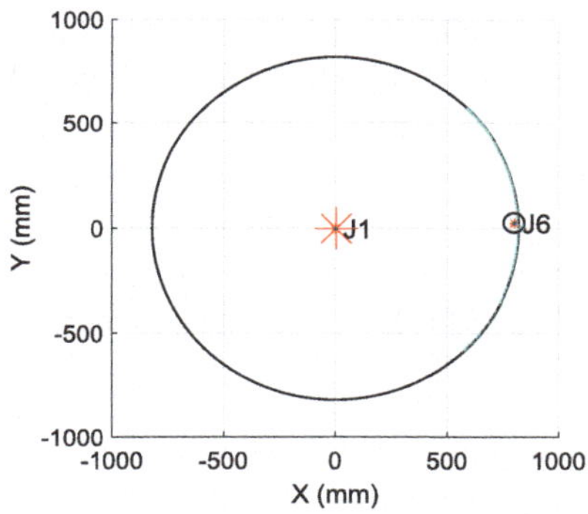

(d) Two views of the circles corresponding to the view of 4.11(c)

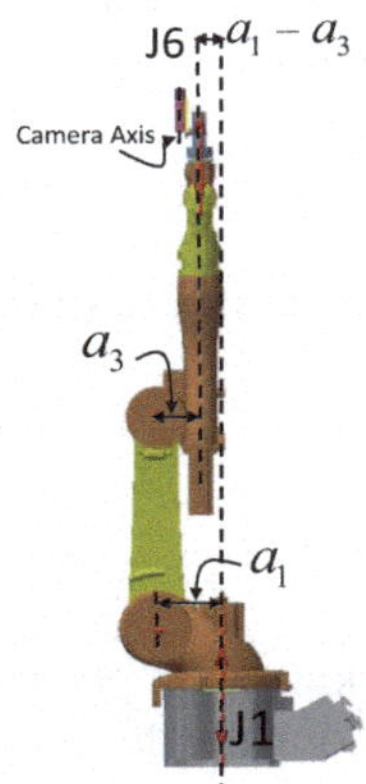

(e) Actuation for joints 1 and 6

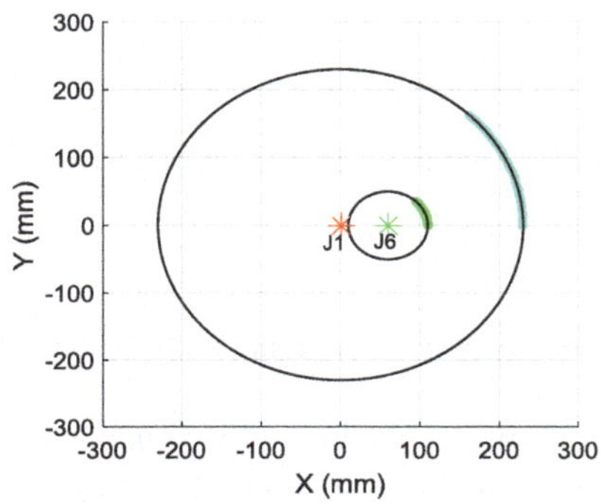

(f) Two views of joints 1 and 6

Fig. 4.11 Motion of end-effector in planes

perpendicular to each another. For the industrial robot KUKA KR5 Arc illustrated in Fig. 4.11, it can be made sure that two joints can be moved such that the EE moves along planes parallel to one another. This is very useful, because of the fact that, the measurements by the camera mounted on the EE can be more accurate if it moves in parallel planes. At first, joints 2, 3, and 5 were actuated such that the camera mounted on the EE moves in parallel planes. The paths moved by the camera mounted on the EE is shown in Fig. 4.11b. Images were captured after which the pose measurements were made. Some parameters/constraints can be obtained directly from the geometry of the robot as shown in Fig. 4.11a.

The link length a_2 for link-2 can be directly obtained as

$$c_{2,3} = a_2. \tag{4.42a}$$

The perpendicular distance between axes 2 and 3 is denoted by $c_{2,3}$. Similarly, the perpendicular distance $c_{3,5}$ between axes 3 and 5, the following constraint relating a_3 (link-length of the third link) and b_4 (joint offset of fourth link) is obtained as

$$c_{3,5} = \sqrt{a_3^2 + b_4^2} \tag{4.42b}$$

Next, with the movement of the EE along the path shown in Fig. 4.11d (configuration is shown in Fig. 4.11c), an additional condition was obtained as

$$c_{1,6} = a_1 + b_4 \tag{4.42c}$$

Combining Eqs. 4.42b and 4.42c a_1 and b_4 can be obtained.

Finally, for the robot configuration in Fig. 4.11e and the actuation of joints 1 and 6, the EE moves along the path as shown in Fig. 4.11f. Given $c'_{1,6}$ the perpendicular distance between axes 1 and 6 as $c'_{1,6}$ a new constraint was obtained that relates a_1 (link-length of first link) and a_3 (link-length of third link) as

$$c'_{1,6} = a_1 - a_3 \tag{4.42d}$$

The value of a_1 obtained above was substituted in Eq. 4.42d to obtain the value of a_3.

4.4.2.1 Discussion on Kinematic Identification Results

Through experiment, an attempt was made to identify the kinematic parameters in the restricted workspace by limited actuation of each joint. Firstly, the comparison between the three measurement devices, namely, TS, LT and monocular camera (MC) is made for the 20° actuation of each joint. Table 4.2 lists the error in length and angular parameters identified with the measurement devices (TS, LT, and MC). The error was quantified by taking Euclidean norm denoted with symbols e_l and e_a.

The error in length and angular parameters e_l and e_a, respectively, for the LT, are much less than that of TS. The reason for such high error in the identified values using the TS is the higher measurement error (SD > 1 mm) associated with the TS. For information about the effect of sensor noise and the errors in estimates, please refer to [71]. Table 4.2 also provides the ratio of the singular values which are inversely proportional to the error e_l and e_a. It can be deduced that the standard deviation (SD) of the measurement noises with TS is greater than 1 mm, i.e., $\Sigma > 1$ mm. With MC, it is between (0.1 and 1) mm and for LT, it is in the order 0.01.

In order to quantify the differences in the identified kinematic parameters w.r.t. the as-desired values in CAD with the root mean square values in Eq. 4.43 for lengths (b and a) and angles (α and θ) as

$$e_l = \sqrt{\frac{\boldsymbol{\epsilon}_{l,ideal} - \boldsymbol{\epsilon}_{l,e}}{\mathtt{dim}(\boldsymbol{\epsilon})}} \text{ and } e_a = \sqrt{\frac{\boldsymbol{\epsilon}_{a,ideal} - \boldsymbol{\epsilon}_{a,e}}{\mathtt{dim}(\boldsymbol{\epsilon})}} \tag{4.43}$$

where $\boldsymbol{\epsilon}_{l,ideal}$ is the vector of the identified length parameters, i.e., $\boldsymbol{\epsilon}_{l,ideal} \equiv [b_1, \cdots, b_6, a_1, \cdots, a_6]^T$ and $\boldsymbol{\epsilon}_{a,ideal}$ is the vector of identified angular parameters as $\boldsymbol{\epsilon}_{a,ideal} \equiv [\theta_1, \cdots, \theta_6, \alpha_1, \cdots, , \alpha_6]^T$ for the CAD model as listed in Table 3.1. The vectors $\boldsymbol{\epsilon}_{l,e}$ and $\boldsymbol{\epsilon}_{a,e}$, on the other hand contain similar data (b, a, θ and α) obtained with the measurement noises in pose data, $\mathtt{dim}(\boldsymbol{\epsilon})$ is the dimension of the vector in the numerator.

Further, the fact observed in simulation that by increasing the sector angle or the joint motions the identified values get closer to the true values. This was experimentally validated by increasing the actuation angle to 50° of each joint using AMM. With 50° sector angle, the ration of σ_1/σ_3 was found to be in the same range as in 20° range, i.e., (177–853). This reflects the fact that the same measurement device, i.e., the Basler camera was used during both the measurements. The error in the identified parameters reduces significantly because of the increased range of joint actuation. It can be easily observed from the Table 4.1 that error in the angular parameters that is mainly the identified twist angle is close to that found with LT.

Unlike [72–75], in this work, an attempt to obtain the geometric parameters of the robot without prior information about its nominal values. Though the identified values of DH parameters using the monocular camera (listed in Table 4.2) with 20° are not close to assumed true values, the results using $\phi = 50°$ showed improvement in the identified values by being closer to true values.

4.4.3 Dynamic Identification of 6-DOF Serial Manipulator

The philosophy of reductionism for analyzing the system by looking at its individual components is vital for identification of a system. In this section, the dynamic identification of a KUKA KR5 Arc robot along with the non-inertial parameters such as kinematic chain, friction, and actuator model.

Table 4.1 DH parameters for KUKA KR5 Arc using measurement devices

	Total Station			Laser Tracker		
Joint	b (mm)	a (mm)	α (Deg.)	b (mm)	a (mm)	α (Deg.)
1	408.87	171.25	89.15	404.29	181.05	90.01
2	0	600.93	179.95	0	600.37	179.81
3	0	110.94	90.20	0	120.11	90.21
4	615.31	10.96	89.50	620.98	1.63	89.94
5	1.56	0.59	90.47	0.56	0	90

Table 4.2 RMS difference in estimated kinematic parameters using measurement devices w.r.t. CAD

MD	SA	σ_1/σ_3	e_l (mm)	e_a (deg)
TS	20°	28–226	36.5	5.6
LT	20°	670–5860	0.45	**0.12**
MV	20°	177–853	7.75	0.50
MV	50°		2.4	**0.08**

MD: Measurement Device, SA:Sector Angle
TS: Total Station, LT: Laser Tracker,
MV: Monocular Vision, e_l and e_a defined in Eq. 4.43

4.4.3.1 Identification of Kinematic Chain

Typically, any robot is driven by motors with a gear box located at the joints or near them. Input to the drives of the motors is current, and output is joint torque. The method for identifying the kinematic chain of KUKA KR5 Arc is presented in this section.

Note that KUKA KR5 Arc is an industrial robot with 6-DOF. The first three joints, namely, 1, 2, and 3 have independent motors. The last three joints, i.e., 4, 5, and 6, are actuated by motors 4, 5, and 6, respectively, as shown in Fig. 4.12, in a coupled manner. The couplings are schematically shown in Fig. 4.13b–c. The last three joints are interconnected such that the motion of the motor 6 affects the motion of motors 4 and 5 in order to compensate the motion of links 4 and 5. The motor current in amperes is denoted by i_i, whereas ε_i and φ_i denote the encoder values and angular displacement of the ith motor, respectively. Moreover, the motor inertia is denoted by $I_{\mathrm{A}i}$, and θ_i is the joint angle. The relationship between the actuator space and the joint space can be represented as

$$\boldsymbol{\varphi} = \mathbf{S}\boldsymbol{\theta} \tag{4.44}$$

where $\boldsymbol{\theta} \equiv [\,\theta_1, \cdots, \theta_6\,]^T$ is the 6D vector of joint angles, whereas $\varphi \equiv [\,\varphi_1, \cdots, \varphi_6\,]^T$ is the 6D vector of the motor shaft rotations. The 6×6 matrix $\mathbf{S}$ is termed as the *structure*

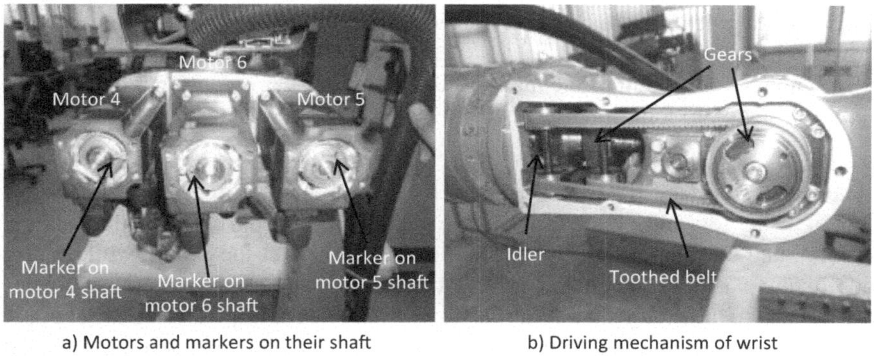

Fig. 4.12 The driving mechanism of KUKA KR5 Arc robot (Please scan QR in the end of this Chapter to see transmission of joints)

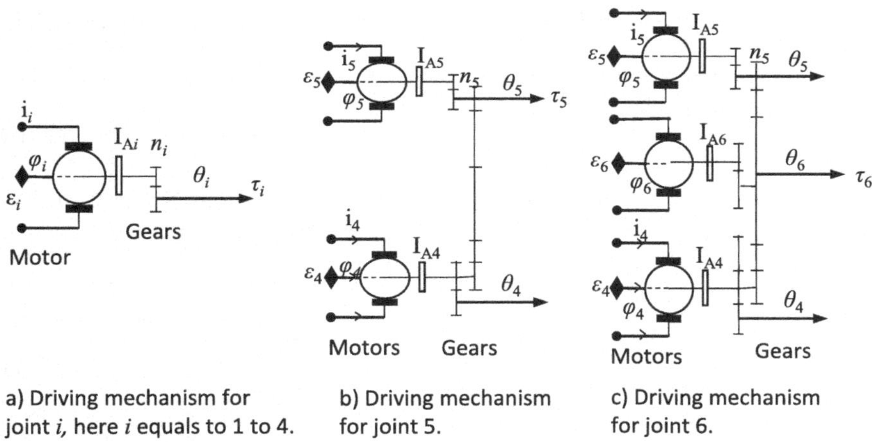

Fig. 4.13 Schematic representation of joint motions of KUKA KR5 Arc robot

matrix relating the joint space to the motor space. The elements of the structure matrix are represented as

$$\mathbf{S} \equiv \begin{bmatrix} s_{11} & \cdots & s_{16} \\ \vdots & \ddots & \vdots \\ s_{61} & \cdots & s_{66} \end{bmatrix} \tag{4.45}$$

The 6×6 matrix **S** in Eq. 4.45 possesses the following properties:

- The structure matrix has full rank, i.e., $rank(S) = 6$. Moreover, the matrix is square. Hence, the determinant must be non-zero, i.e., $det(S) \neq 0$.

- Non-zero elements in the off-diagonal terms represent the dependence of one joint on another.
- Each non-zero element of the matrix **S** indicates the gear reduction ratio from the motor rotation to joint rotation.

In order to identify the elements of matrix **S** for KUKA KR5 Arc, each joint was rotated one by one starting from the first to the sixth. Figure 4.13c shows the arrangement of the motors and encoders using line diagram mounted for the last three joints, i.e., 4, 5, and 6, on the third link. Sixth motor is located at the center, while fourth and fifth are placed to the left and right of it, respectively. The encoder counts are linearly related to the motor shaft rotation. To find the relationship between the encoder increment and the rotation of the motor shaft, the complete rotation, i.e., by 360 degrees was provided to the motor shaft and the encoder counts were noted from Robot Sensory Interface (RSI). The full rotation was checked by the markers placed on the motor shafts as shown in Fig. 4.12a. The joints meant for the wrist movement of the robot are driven a combination of twisted toothed belts with an idler and gears shown in Fig. 4.12b.

The variation of motor rotations (φ_i's) in degrees and the linear varying encoder counts (ε_i's) can be related as

$$\boldsymbol{\varphi} = \mathbf{T}_m \boldsymbol{\varepsilon} \tag{4.46}$$

where, $\boldsymbol{\varphi} \equiv [\, \varphi_1, \cdots, \varphi_6 \,]^T$, is the 6D vector of the motor rotations, $\mathbf{T}_m$ is the 6×6 constant diagonal matrix and $\boldsymbol{\varepsilon} \equiv [\, \varepsilon_1, \cdots, \varepsilon_6 \,]^T$ was the 6D vector of encoder counts. Matrix $\mathbf{T}_m$ is obtained as

$$\mathbf{T}_m = 360 \begin{bmatrix} \begin{matrix} 1/16390 & 0 & 0 \\ 0 & 1/16390 & 0 \\ 0 & 0 & 1/16386 \end{matrix} & \mathbf{O}_{3\times3} \\ \mathbf{O}_{3\times3} & \begin{matrix} 1/12300 & 0 & 0 \\ 0 & 1/12833 & 0 \\ 0 & 0 & 1/12844 \end{matrix} \end{bmatrix} \tag{4.47}$$

where $\mathbf{O}_{3\times3}$ imply zeros of the off-diagonal elements. In Eq. (4.47), the denominators were obtained by taking the difference of the encoders count after each complete rotation and averaging it for the readings under encoder counts columns in Table B.2. Next, to relate the encoder counts with joint readings, some simultaneous joint movement commands were given to the robot's controller. Corresponding encoder counts were recorded.

$$\boldsymbol{\epsilon} = \mathbf{T}_j \boldsymbol{\theta} \tag{4.48}$$

where $\mathbf{T}_j$ is a 6×6 constant matrix. Using the values of Table B.3, the elements of 6×6 matrix $\mathbf{T}_j$ can be obtained as

$$\mathbf{T}_j = \begin{bmatrix} \begin{matrix} 5.68 & 0 & 0 \\ 0 & 5.68 & 0 \\ 0 & 0 & 3.95 \end{matrix} & \mathbf{O}_{3\times3} \\ \mathbf{O}_{3\times3} & \begin{matrix} 2.54 & 0 & 0 \\ -0.03 & 1.43 & 0 \\ -0.03 & 0.03 & 0.82 \end{matrix} \end{bmatrix} \tag{4.49}$$

In Eq. (4.49), first four diagonal elements were found using the readings of Table B.3 and noting the independent joint relation as

$$\varepsilon_1 = t_{11}\theta_1,\ \varepsilon_2 = t_{22}\theta_2,\ \varepsilon_3 = t_{33}\theta_3,\ \text{and}\ \varepsilon_4 = t_{44}\theta_4 \tag{4.50}$$

$$\varepsilon_5 = t_{54}\theta_4 + t_{55}\theta_5 \tag{4.51}$$

and

$$\varepsilon_6 = t_{64}\theta_4 + t_{65}\theta_5 + t_{66}\theta_6 \tag{4.52}$$

where t_{ii}, for $i = 1, \cdots, 4$ are the first four diagonal elements of the matrix $\mathbf{T}_j$. The other non-zero elements of $\mathbf{T}_j$ in Eq. (4.49) were evaluated using the readings of Table B.3 and for the coupled relation for wrist action, t_{54}, t_{55}, t_{64}, t_{65} and t_{66} are the non-zero coefficients of $\mathbf{T}_j$, which were found by substituting values from Table B.3 in Eq. (4.52) and taking generalized inverse of the associated coefficient matrix. On combining Eqs. (4.46) and (4.49), $\boldsymbol{\varphi}$ is obtained as

$$\boldsymbol{\varphi} = \mathbf{S}\boldsymbol{\theta} \tag{4.53}$$

where $\mathbf{S} = \mathbf{T}_m\mathbf{T}_j$. The 6×6 matrix $\mathbf{S}$ for KUKA KR5 Arc robot is then found as

$$\mathbf{S} = \begin{bmatrix} 125.04 & 0 & 0 & & & \\ 0 & 125.05 & 0 & & \mathbf{O}_{3\times3} & \\ 0 & 0 & 89.68 & & & \\ & & & 74.59 & 0 & 0 \\ & \mathbf{O}_{3\times3} & & -1.00 & 42.00 & 0 \\ & & & -1.00 & 1.00 & 24.43 \end{bmatrix} \tag{4.54}$$

The magnitudes of non-zero elements in the first four rows of $\mathbf{S}$ in Eq. (4.54) give the gear reduction ratio for the first four joints of the robot. Similarly, the sixth row with three non-zero elements represents the coupling of joint 6 by motors 4, 5, and 6. The fifth row with two non-zero elements represents the coupling effect, i.e., joint 5 is driven by motors 4 and 5. The negative sign represents the motion is coupled with an opposite rotation of the motors. For example, the element $s_{64} = -1$ of Eq. (4.54) represents that the fourth motor will rotate in the opposite direction to provide rotation to the 5th joint. Equations (4.53) and (4.54) are important in identifying the kinematic chain of the KUKA KR5 Arc robot.

4.4.3.2 Data Acquisition, Processing, and Filtering

The joint readings of KUKA KR5 Arc robot were taken from joint encoders and the joint current values were taken from the sensor attached to each joint. The robot was communicated through TCP/IP using `ST_ETHERNET` command of KUKA RSI 2.3 installed on KUKA KRC2 robot controller. The data were sampled at 12 ms. The joint velocity and acceleration were obtained using central difference algorithm.

For removing the noise in the joint position data, zero-phase digital filtering through an IIR low-pass butterworth filter in both the forward and reverse direction was done using a `filtfilt(b,a,x)` command of MATLAB®. The co-efficients

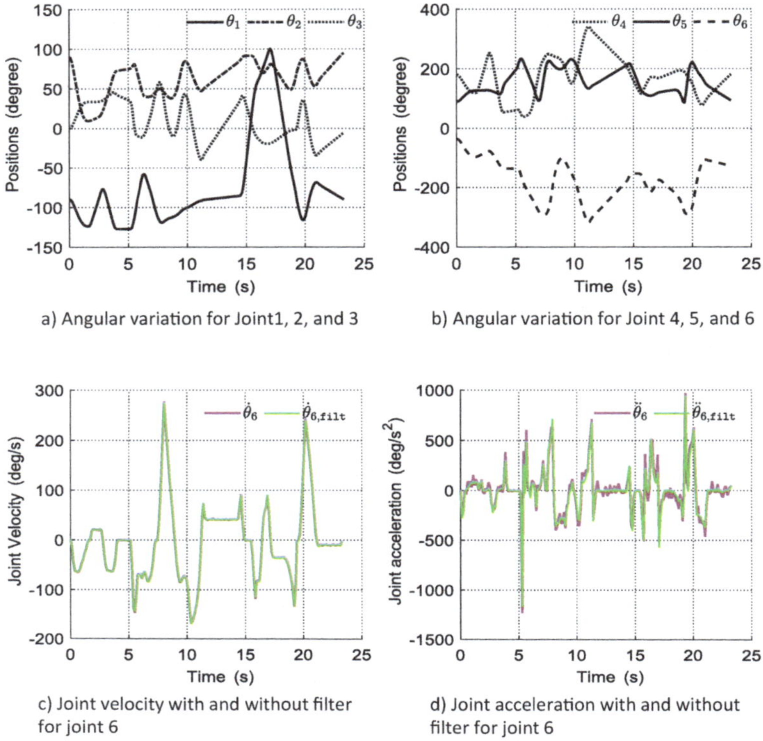

Fig. 4.14 Joint trajectories used for the identification of base parameters

`b` and `a` were found using second-order Butterworth filter `butter(n,Wn)` with order of filter `n=2` and normalized cutoff frequency `Wn` as (7/41.66) where, 41.66 in the denominator is the Nyquist frequency, i.e., half of the sampling frequency and 7 Hz is the signal cutoff frequency. The excitation trajectory was given to the robot using point-to-point (PTP) motion with a total of 19520 data points decimated by a factor of 10. By default, `decimate` is a command in `MATLAB`® which uses a low-pass Chebyshev Type I IIR filter of order 8 [76]. Whereas Fig. 4.14a, b show variations in joint angle, Fig. 4.14c, d show the velocity and acceleration plot for joint 6 before and after applying noise filter.

4.4.4 Modeling and Parameter Estimation

The frictional and actuator model were taken into account for the installed robot which was not considered in simulation. The processed 1948 data consisting of joint trajectory and torque obtained after decimation in Sect. 4.4.3.2 were concatenated to get observation matrix as in Eq. (4.39). The condition number of $\mathbf{Y}_{b,\mathrm{T}}$ for the trajectory used (angular variation is shown in Fig. 4.14) for identifying the BP was 51. Note that this condition number was easily achieved by exciting the robot with a PTP motion with sudden change in positions and directions. These are evident from the joint variation plot shown in Fig. 4.14a, b. The position vector of the EE w.r.t the origin are plotted in Fig. 4.14c, d. The vector of BPs, namely, χ was estimated using Eq. (4.41). Table 4.3 lists the identified BP for the installed manipulator. Note that out of 78 (13×6) parameters for the six links which gets reduced to only 52 base parameters (BPs).

4.4.4.1 Results and Discussions

The identified model was validated by comparing the model joint torques calculated using the identified model with those measured from the robot controller for a given trajectory as shown in Fig. 4.15b which was different from the identification trajectory as shown in Fig. 4.15a. The torque reconstruction results are shown in Fig. 4.16 for the same trajectoyr used for the identification. As observed from the plots the joint torques are in the close match. Table 4.4 lists the percentage RMS errors between the measured and the predicted toques.

Note that some of the rows in Table 4.3 were highlighted in shades of gray color to point out that these BPs were having significantly larger values compared to the rest. The larger values for the highlighted BPs are either because of the fact that these BPs contains a large number of linearly combined SIP or individually influence the dynamics. For example, the viscous and Coulomb friction coefficients of joint 1, 2, and 3 are having higher numeric values compared to other base parameters. Interestingly, the coefficient of viscous friction is higher than the Coulomb friction for these joints. The reason for this is that coefficient of friction is greatly affected by torque level, position, velocity, acceleration, and lubrication properties, etc. Particularly, the viscous friction effect increases with the increase in velocity.

Figure 4.17 depicts the torque reconstruction with the identified BPs and the test trajectory shown Fig. 4.15b. It is evident from the torque plots of the six joints at Fig. 4.17, the magnitude of the torques in joints 1, 2, and 3 are much higher when the robot is running at operational velocity and sudden variation in the direction of the motion, i.e., position vector of the EE that can be perceived from Fig. 4.15d. For example, just before the time stamp of 10 seconds in all the joints torque in Fig. 4.17 shows the peak which is due to the sudden change in position vector at the same time stamp shown in Fig. 4.15b. Thanks to the persistent excitation trajectory which resulted in the identification with fairly good torque reconstruction. Identification of the kinematic chain and addition of the friction model in the dynamics led to achieving this. Table 4.4 lists the RMS errors and percentage RMS errors between the torque reconstructed using the identified model and the measured torques for identification and the test trajectories.

Table 4.3 Symbolic expressions of the BPs and their numeric values

	Base Parameters expression (BPs χ_b)	Values
1	$I_{1a}+k_1+0.0324m_1-0.309375m_2-0.2691\bar{m}_3+0.36m_1d_{1x}+0.54m_3d_{3y}+0.27m_2d_{2z}$	4.8631
2	F_{1c}	54.6099
3	F_{1v}	212.0953
4	$I_{2xx}-I_{2yy}+0.36\bar{m}_2$	16.7731
5	I_{2xy}	0.7086
6	$I_{2xz}-0.081\bar{m}_3-0.6m_3d_{3y}-0.6m_2d_{2z}$	2.2833
7	I_{2yz}	2.114
8	$I_{2a}+I_{2zz}-0.36\bar{m}_2$	–0.8037
9	$0.6\bar{m}_2+m_2d_{2x}$	27.9157
10	m_2d_{2y}	0.392
11	F_{2c}	63.6172
12	F_{2v}	249.9446
13	$I_{3xx}-I_{3zz}+I_{4zz}+0.0144m_3+0.3988\bar{m}_4+1.24m_4d_{4y}$	3.1623
14	$I_{3xy}-0.12m_3d_{3y}$	1.7449
15	I_{3xz}	1.3157
16	$I_{3yy}+I_{4zz}-0.0144m_3+0.37\bar{m}_4+1.24m_4d_{4y}$	4.4695
17	I_{3yz}	1.6444
18	$0.12\bar{m}_3+m_3d_{3x}$	4.8002
19	$0.62\bar{m}_4+m_4d_{4y}+m_3d_{3z}$	6.3216
20	I_{3a}	3.2538
21	F_{3c}	23.1787
22	F_{3v}	70.0084
23	$I_{4xx}-I_{4zz}+I_{5zz}$	–0.2108
24	I_{4xy}	–0.0871
25	I_{4xz}	0.0481
26	$I_{4yy}+I_{5zz}$	–0.6187
27	I_{4yz}	–0.5547
28	m_4d_{4x}	–0.0081
29	$m_5d_{5y}+m_4d_{4z}$	–0.1468
30	I_{4a}	2.7173
31	F_{4c}	15.0044
32	F_{4v}	15.8814
33	$I_{5xx}+I_{6yy}-I_{5zz}+0.013225m_6+0.23m_6d_{6z}$	–0.3523
34	I_{5xy}	0.0503
35	I_{5xz}	–0.2269
36	$I_{5yy}+I_{6yy}+0.013225m_6+0.23m_6d_{6z}$	–0.2374

(continued)

Table 4.3 (continued)

	Base Parameters expression (BPs χ_b)	Values
37	I_{5yz}	0.0541
38	m_5d_{5x}	0.0247
39	$0.115m_6 + m_5d_{5z} + m_6d_{6z}$	0.2817
40	I_{5a}	0.7804
41	F_{5c}	5.3391
42	F_{5v}	3.5852
43	$I_{5xx} - I_{6yy}$	0.2026
44	I_{6xy}	–0.015
45	I_{6xz}	–0.0087
46	I_{6yz}	0.0516
47	I_{6zz}	–0.1736
48	m_6d_{6x}	0.2146
49	m_6d_{6y}	0.1191
50	I_{6a}	0.1831
51	F_{6c}	2.4536
52	F_{6v}	4.3397

Note that symbol $\bar{m}_i = \sum_{j=1}^{i} m_j$

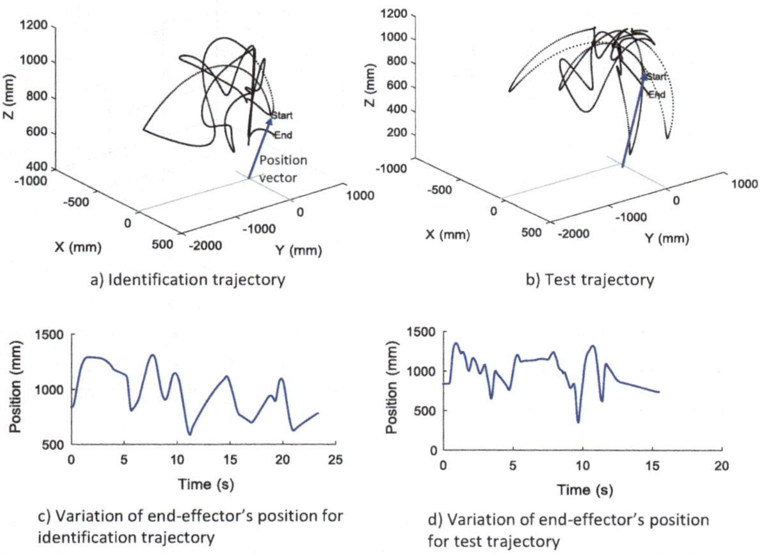

Fig. 4.15 Trace of identification and test trajectory given while experiment

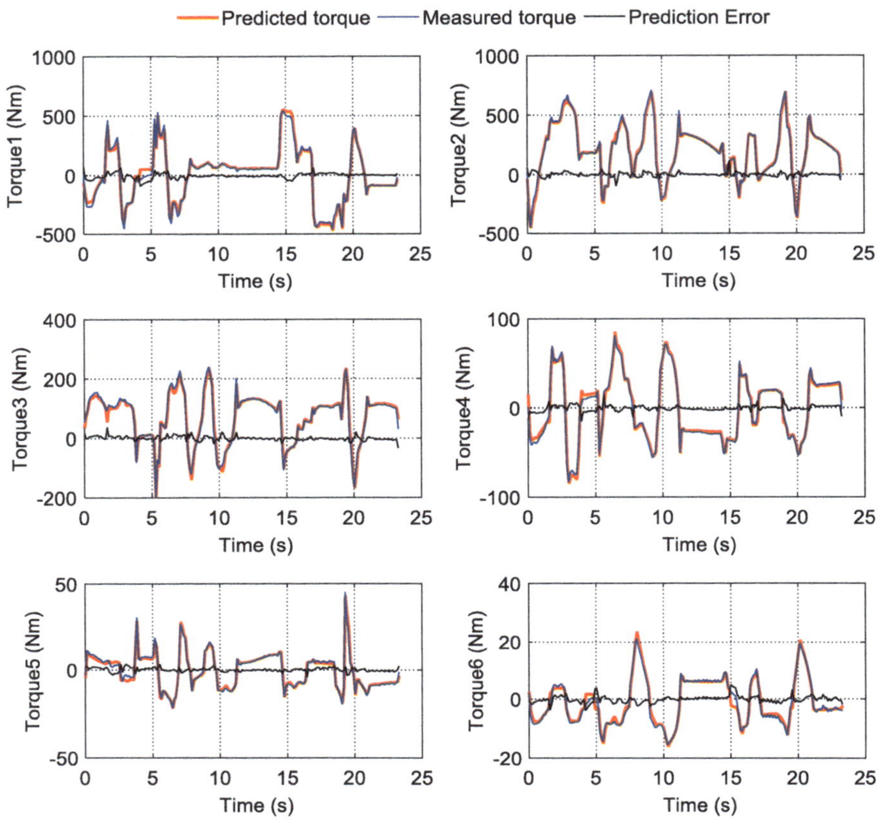

Fig. 4.16 Comparison between measured and estimated torques using identification trajectory (Please scan QR code in the end of this Chapter to see video of the trajectory)

Table 4.4 RMS error in the torque reconstruction using identification and test trajectory

Trajectory ↓	Joints →	1	2	3	4	5	6
Identification	RMS error	23.78	20.17	8.41	3.29	1.32	1.22
	% RMS error	10.20	6.55	7.61	8.91	12.79	15.31
Test	RMS error	17.94	27.05	13.06	6.03	3.83	2.85
	% RMS error	11.60	14.36	12.20	21.87	31.87	37.80

The advantage of the accurate identification model is in torque estimation for simulation and control. The simplified model were identified for the robot performing tasks at very slow speed, i.e., $\dot{\theta}_i, \ddot{\theta}_i$ are negligible [41]. The gravitational torque and the torque required to overcome the friction were mainly estimated by the model identified in the work coauthored in [41]. With the identified model, the current required to maintain the position were predicted accurately and limiting the current

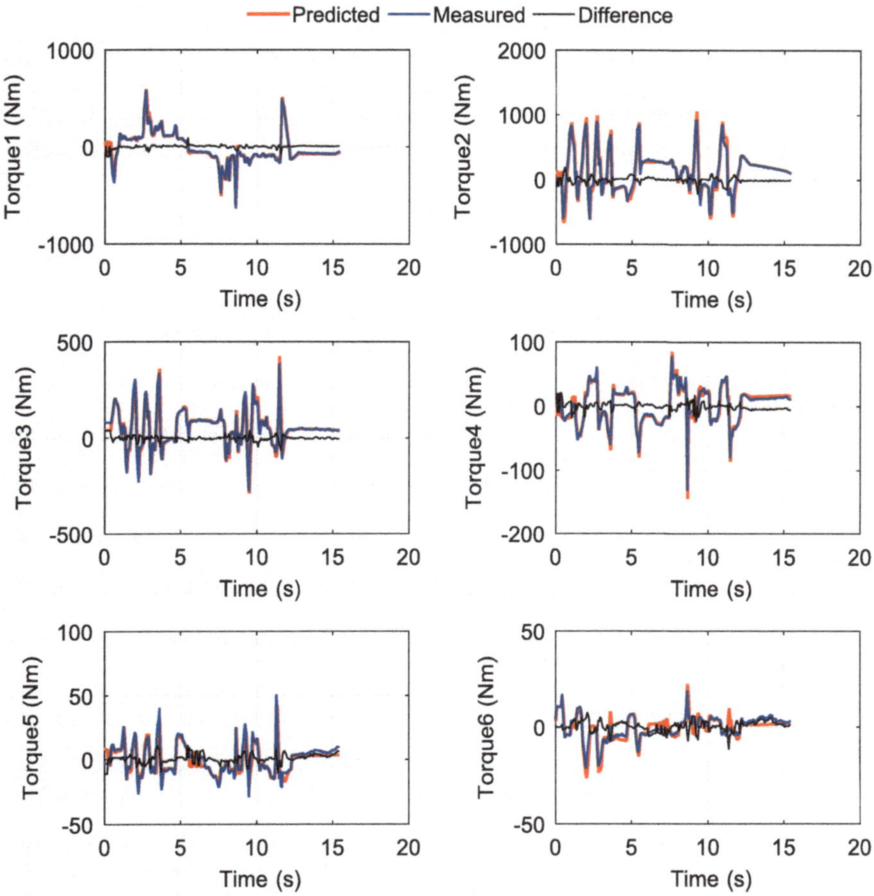

Fig. 4.17 Comparison between measured and estimated torques using the test trajectory

approach resulted in the industrial robot behave like a passively compliant. Limiting current method implemented in the controller with the identified model may result in intrinsically safe from potential pinching or trapping hazards from the robot, and will also minimize the collision hazards.

4.5 Summary

In this chapter, firstly, kinematic identification of a robot is proposed based on circle point method (CPM) and Singular Value Decomposition (SVD). The ratio of largest to smallest singular values gives a clear indication about the quality quality of identified parameters. The use of a monocular camera was emphasized for the identification purposes. Secondly, the formulation for the linear inverse dynamics identification model is proposed using the DeNOC matrices. The experimental identification of the installed KUKA KR5 Arc robot was presented including the effect of friction and transmission system. The identified models were validated as well.

References

1. Denavit, J., Hartenberg, R.S.: A kinematic notation for lower-pair mechanisms based on matrices. Trans. ASME J. Appl. Mech. **22**, 215–221 (1955)
2. Santolaria, J., Ginés, M.: Uncertainty estimation in robot kinematic calibration. Robot. Comput. Integr. Manuf. **29**(2), 370–384 (2013)
3. Xiao, W., Huan, J., Dong, S.: A step-compliant industrial robot data model for robot off-line programming systems. Robot. Comput. Integr. Manuf. **30**(2), 114–123 (2014)
4. Khalil, W., Dombre, E.: Modeling, Identification and Control of Robots. Butterworth-Heinemann (2004)
5. Roth, Z.S., Mooring, B., Ravani, B.: An overview of robot calibration. IEEE J. Robot. Automat. **3**(5), 377–385 (1987)
6. Benjamin, M., Zvi, S.R., Morris, R.D.: Fundamentals of Manipulator Calibration. Wiley, New York (1991)
7. Bugalia, N., Sen, A., Kalra, P., Kumar, S.: Immersive environment for robotic tele-operation. In: Proceedings of Advances in Robotics, pp. 1–6. ACM (2015)
8. Mukherjee, S., Zubair, M., Suthar, B., Kansal, S.: Exoskeleton for tele-operation of industrial robot. In: Proceedings of Conference on Advances In Robotics, AIR, pp. 108:1–108:5. New York, USA (2013)
9. Bennett, D.J., Hollerbach, J.M.: Autonomous calibration of single-loop closed kinematic chains formed by manipulators with passive endpoint constraints. IEEE Trans. Robot. Automat. **7**(5), 597–606 (1991)
10. Jagadeesh, B., IssacKurien, K.: Calibration of a closed loop three degree of freedom manipulator. In: Proceedings of the 6th World Congress of IFToMM, pp. 1762–1766. IFToMM (1995)
11. Clark, L., Shirinzadeh, B., Tian, Y., Fatikow, S., Zhang, D.: Pose estimation and calibration using nonlinear capacitance sensor models for micro/nano positioning. Sens. Actuators Phys. **253**, 118–130 (2017)
12. Escande, C., Chettibi, T., Merzouki, R., Coelen, V., Pathak, P.M.: Kinematic calibration of a multisection bionic manipulator. IEEE/ASME Trans. Mechatron. **20**(2), 663–674 (2015)
13. Abderrahim, M., Whittaker, A.: Kinematic model identification of industrial manipulators. Robot. Comput. Integr. Manuf. **16**(1), 1–8 (2000)
14. Hayat, A.A., Elangovan, K., Rajesh Elara, M., Teja, M.S.: Tarantula: Design, modeling, and kinematic identification of a quadruped wheeled robot. Appl. Sci. **9**(1), 94 (2019)
15. Boby, R.A., Klimchik, A.: Combination of geometric and parametric approaches for kinematic identification of an industrial robot. Robot. Comput. Integr. Manuf. **71**, 102142 (2021)
16. Chiwande, S.N., Ohol, S.S.: Comparative need analysis of industrial robot calibration methodologies. In: IOP Conference Series: Materials Science and Engineering, vol. 1012, pp. 012009. IOP Publishing (2021)
17. Kwon, J., Choi, K., Park, F.C.: Kinodynamic model identification: A unified geometric approach. IEEE Trans. Robot. (2021)
18. Huston, R., Passerello, C.: On constraint equations a new approach. J. Appl. Mech. **41**(4), 1130–1131 (1974)
19. Walker, M.W., Orin, D.E.: Efficient dynamic computer simulation of robotic mechanisms. J. Dyn. Syst. Meas. Control **104**(3), 205–211 (1982)

20. Kim, S., Vanderploeg, M.: A general and efficient method for dynamic analysis of mechanical systems using velocity transformations. J. Mech. Trans. Automat. Des. **108**(2), 176–182 (1986)
21. Hemami, H., Weimer, F.: Modeling of nonholonomic dynamic systems with applications. J. Appl. Mech. **48**(1), 177–182 (1981)
22. Mani, N., Haug, E., Atkinson, K.: Application of singular value decomposition for analysis of mechanical system dynamics. J. Mech. Trans. Automat. Des. **107**(1), 82–87 (1985)
23. Kim, S., Vanderploeg, M.: QR decomposition for state space representation of constrained mechanical dynamic systems. J. Mech. Trans. Automat. Des. **108**(2), 183–188 (1986)
24. Angeles, J., Lee, S.K.: The formulation of dynamical equations of holonomic mechanical systems using a natural orthogonal complement. J. Appl. Mech. **55**(1), 243–244 (1988)
25. Saha, S.K., Angeles, J.: Dynamics of nonholonomic mechanical systems using a natural orthogonal complement. J. Appl. Mech. **58**(1), 238–243 (1991)
26. Cyril, X.: Dynamics of Flexible Link Manipulators. PhD thesis, McGill University (1988)
27. Saha, S.K.: Analytical expression for the inverted inertia matrix of serial robots. Int. J. Robot. Res. **18**(1), 116–124 (1999)
28. Gupta, V., Saha, S.K., Chaudhary, H.: Optimum design of serial robots. J. Mech. Des. **141**(8) (2019)
29. Hayat, A.A., Saha, S.K.: Identification of robot dynamic parameters based on equimomental systems. In: Proceedings of the Advances in Robotics, pp. 1–6 (2017)
30. Swevers, J., Verdonck, W., De Schutter, J.: Dynamic model identification for industrial robots. IEEE Control Syst. **27**(5), 58–71 (2007)
31. Atkeson, C., An, C., Hollerbach, J.: Rigid body load identification for manipulators. In: 24th IEEE Conference on Decision and Control, vol. 24, pp. 996–1002 (1985)
32. Atkeson, C.G., An, C.H., Hollerbach, J.M.: Estimation of inertial parameters of manipulator loads and links. Int. J. Robot. Res. **5**(3), 101–119 (1986)
33. Khosla, P.K., Kanade, T.: Parameter identification of robot dynamics. In: 1985 24th IEEE Conference on Decision and Control, pp. 1754–1760. IEEE (1985)
34. Khalil, W., Kleinfinger, J.-F.: Minimum operations and minimum parameters of the dynamic models of tree structure robots. IEEE J. Robot. Automat. **3**(6), 517–526 (1987)
35. Gautier, M., Khalil, W.: Direct calculation of minimum set of inertial parameters of serial robots. IEEE Trans. Robot. Automat. **6**(3), 368–373 (1990)
36. Mata, V., Benimeli, F., Farhat, N., Valera, A.: Dynamic parameter identification in industrial robots considering physical feasibility. Adv. Robot. **19**(1), 101–119 (2005)
37. Farhat, N., Mata, V., Page, A., Valero, F.: Identification of dynamic parameters of a 3-DOF RPS parallel manipulator. Mech. Mach. Theory **43**(1), 1–17 (2008)
38. Jain, A.: Robot and Multibody Dynamics: Analysis and Algorithms. Springer Science & Business Media (2010)
39. Le Tien, L., Schaffer, A.A., Hirzinger, G.: Mimo state feedback controller for a flexible joint robot with strong joint coupling. In: International Conference on Robotics and Automation, pp. 3824–3830. IEEE (2007)
40. Waiboer, R., Aarts, R., Jonker, J.: Application of a perturbation method for realistic dynamic simulation of industrial robots. Multibody Syst. Dyn. **13**(3), 323–338 (2005)
41. Udai, A.D., Hayat, A.A., Saha, S.K.: Parallel active/passive force control of industrial robots with joint compliance. In: IEEE/RSJ International Conference on Intelligent Robots and Systems, pp. 4511–4516. IEEE (2014a)
42. Armstrong-Hélouvry, B.: Control of Machines with Friction, vol. 128. Springer Science & Business Media (2012)
43. Armstrong-Hélouvry, B., Dupont, P., De Wit, C.C.: A survey of models, analysis tools and compensation methods for the control of machines with friction. Automatica **30**(7), 1083–1138 (1994)
44. Dupont, P.E.: Friction modeling in dynamic robot simulation. In: IEEE International Conference on Robotics and Automation, pp. 1370–1376. IEEE (1990)
45. Grotjahn, M., Daemi, M., Heimann, B.: Friction and rigid body identification of robot dynamics. Int. J. Solids Struct. **38**(10), 1889–1902 (2001)

46. Grotjahn, M., Heimann, B., Abdellatif, H.: Identification of friction and rigid-body dynamics of parallel kinematic structures for model-based control. Multibody Syst. Dyn. **11**(3), 273–294 (2004)
47. Swevers, J., Al-Bender, F., Ganseman, C.G., Projogo, T.: An integrated friction model structure with improved presliding behavior for accurate friction compensation. IEEE Trans. Automat. Control **45**(4), 675–686 (2000)
48. Armstrong, B.: On finding exciting trajectories for identification experiments involving systems with nonlinear dynamics. Int. J. Robot. Res. **8**(6), 28–48 (1989)
49. Gautier, M., Khalil, W.: Exciting trajectories for the identification of base inertial parameters of robots. Int. J. Robot. Res. **11**(4), 362–375 (1992)
50. Swevers, J., Ganseman, C., Tukel, D.B., De Schutter, J., Van Brussel, H.: Optimal robot excitation and identification. IEEE Trans. Robot. Automat. **13**(5), 730–740 (1997)
51. Calafiore, G., Indri, M., Bona, B.: Robot dynamic calibration: Optimal excitation trajectories and experimental parameter estimation. J. Robot. Syst. **18**(2), 55–68 (2001)
52. Jin, J., Gans, N.: Parameter identification for industrial robots with a fast and robust trajectory design approach. Robot. Comput. Integr. Manuf. **31**, 21–29 (2015)
53. Serban, R., Freeman, J.: Identification and identifiability of unknown parameters in multibody dynamic systems. Multibody Syst. Dyn. **5**(4), 335–350 (2001)
54. Nelles, O.: Nonlinear System Identification: From Classical Approaches to Neural Networks and Fuzzy Models. Springer Science & Business Media (2013)
55. Hayat, A. A., Sadanand, R. O., Saha, S. K.: Robot manipulation through inverse kinematics. In Proceedings of the 2015 Conference on Advances In Robotics, 1–6 (2015)
56. Golub, G.H., Van Loan, C.F.: Matrix Computations, vol. 3. JHU Press (2012)
57. Pennestrì, E., Stefanelli, R.: Linear algebra and numerical algorithms using dual numbers. Multibody Syst. Dyn. **18**(3), 323–344 (2007)
58. Othayoth, R.S., Chittawadigi, R.G., Joshi, R.P., Saha, S.K.: Robot kinematics made easy using roboanalyzer software. Comput. Appl. Eng. Educ. (2017)
59. Saha, S.K.: Introduction to Robotics, 2nd edn. Tata McGraw-Hill Education (2014b)
60. Saha, S.K., Schiehlen, W.O.: Recursive kinematics and dynamics for parallel structured closed-loop multibody systems. Mech. Struct. Mach. **29**(2), 143–175 (2001)
61. Angeles, J., Lee, S.: The formulation of dynamical equations of holonomic mechanical systems using a natural orthogonal complement. ASME J. Appl. Mech. **55**, 243–244 (1988)
62. Saha, S.K.: A decomposition of the manipulator inertia matrix. IEEE Trans. Robot. Automat. **13**(2), 301–304 (1997)
63. Pauline, H., Gautier, M., Phillippe, G.: Dynamic identification of robots with a dry friction model depending on load and velocity. In: IEEE International Conference on Robotics and Automation, pp. 6187–6193. IEEE (2010)
64. Gautier, M.: Numerical calculation of the base inertial parameters of robots. J. Robot. Syst. **8**(4), 485–506 (1991)
65. Strang, G.: Introduction to Linear Algebra. Wellesley-Cambridge Press (2009)
66. OpenCV: Open Source Computer Vision (2020). http://opencv.org/. [Online; accessed 13-October 2020]
67. Zhang, Z.: A flexible new technique for camera calibration. IEEE Trans. Pattern Anal. Mach. Intell. **22**, 1330–1334 (2000)
68. ArUco: ArUco: a minimal library for augmented reality applications based on opencv (2020). ArUco: a minimal library for Augmented Reality applications based on OpenCV. [Online; accessed 18-December-2020]
69. Rafael, M., Manuel, M., Enrique, Y., Rafael, M.: Mapping and localization from planar markers. Under Rev IEEE Trans. Pattern Anal. Mach. Intell. (2016). arXiv:1606.00151
70. OpenCV: Camera Calibration and 3D Reconstruction (2020). http://docs.opencv.org/2.4/modules/calib3d/doc/camera_calibration_and_3d_reconstruction.html. [Online; accessed 06-October-20]
71. Hayat, A.A., Boby, R.A., Saha, S.K.: A geometric approach for kinematic identification of an industrial robot using a monocular camera. Robot. Comput. Integr. Manuf. **57**, 329–346 (2019)

72. Rousseau, P., Desrochers, A., Krouglicof, N.: Machine vision system for the automatic identification of robot kinematic parameters. IEEE Trans. Robot. Automat. **17**(6), 972–978 (2001)
73. Meng, Y., Zhuang, H.: Self-calibration of camera-equipped robot manipulators. Int. J. Robot. Res. **20**, 909–921 (2001)
74. Motta, J.M.S., de Carvalho, G.C., McMaste, R.: Robot calibration using a 3D vision-based measurement system with a single camera. Robot. Comput. Integr. Manuf. **17**, 487–497 (2001)
75. Du, G., Zhang, P.: Online robot calibration based on vision measurement. Robot. Comput. Integr. Manuf. **29**(6), 484–492 (2013)
76. MATLAB: http://in.mathworks.com/help/signal/filter-responses-and-design-methods.html (2014). [Online; accessed 20-September-2016]

Chapter 5
Force Control and Assembly

This chapter explains the force control strategy used for the parallel active and passive force control of an industrial manipulator. The approach presented is a generic way and is useful for the safer assembly task. This also makes the robot compliant once the obstruction imposes a reaction force greater than the one obtained based on the identified dynamic model as detailed in previous chapter. A simplified approach is proposed to estimate the dynamic parameters which are useful for the constraint task of peg-in-tube. Force control becomes a prerequisite when it comes to performing any contact-based task using a robot so that the inaccuracy in the robot's end-effector position or any uncertain changes in its environment could be dealt with ease. The detailed approach for the assembly task, namely, peg-in-tube is discussed.

5.1 Background

The background study is done in two steps, namely, the force control strategies presented in literature in context to transform the robot safer for the interaction purpose with the environment or human. Next, the assembly task focused at peg-in-tube using an industrial robot is highlighted.

Supplementary Information The online version contains supplementary material available at https://doi.org/10.1007/978-981-16-6990-3_5.

A. A. Hayat et al., *Vision Based Identification and Force Control of Industrial Robots*,
Studies in Systems, Decision and Control 404,
https://doi.org/10.1007/978-981-16-6990-3_5

5.1.1 Force control

Figure 5.1a shows a cooperative manipulation, where one of the robots leads the desired path and the second one follows the trajectory based on the forces imposed by the first one. Similarly, Fig. 5.1b shows a peg-in-hole task being performed by an industrial robot, where the end effector is required to stop when it makes contact with the surface. Once the hole is detected during insertion the end effector is free to move along the vertical hole direction, but constrained along remaining directions. During these stated tasks, the end effector is required to move on a geometrically constrained path. The robot is also expected to sense any external force and take corrective measures to finally execute the desired command. Having a torque control at the joints has been proven to be instrumental in designing force control algorithms for research robots. A good number of these approaches like stiffness, impedance, admittance, and hybrid controls have been discussed by [1, 2], and a variety of such robots were commercialized for domestic environment. However, these robots are not common in the industry due to its low precision, smaller payload capacity, less ruggedness, and high initial capital cost. An industrial setup with open control architecture is considered paradigm for the implementation of such interaction control [3]. However, any industrial robot manufacturer restricts the access to robot's intrinsic control parameters for its joint servos, controller gains, dynamic constants like mass moment of inertia and link masses, etc., probably to maintain its finely tuned safety and quality standards. Industrial robots are powerful and are meant to perform precise tasks with strict quality standards. Any mishandling may cause immense damage to the environment and property. These restrictions have proven to be major hindrance toward the development of robots for specialized applications like compliant manipulation and interaction control or more formally for *force control*. This resulted in a very few literature on force control of industrial robots.

Schutter and Brussel [4] suggested a controller design which was based on an *external force control loop* that was closed around the robot's existing positioning

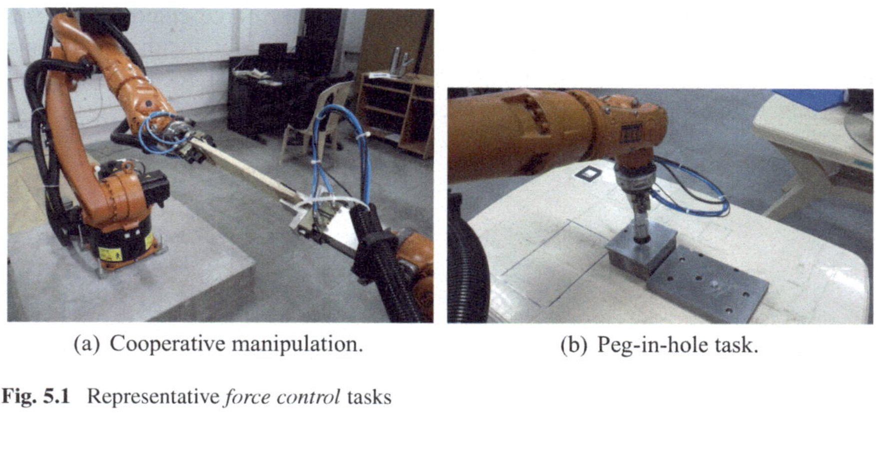

(a) Cooperative manipulation. (b) Peg-in-hole task.

Fig. 5.1 Representative *force control* tasks

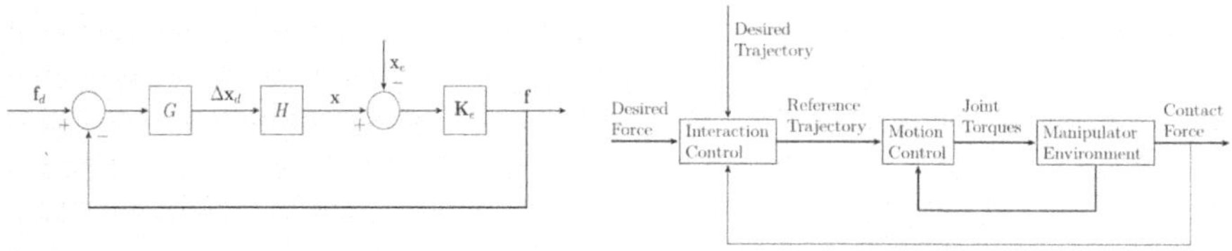

(a) External force control loop (b) Interaction control scheme

Fig. 5.2 Control schemes implemented in industrial manipulators

system. In Fig. 5.2a, $\mathbf{f}_d$ and $\mathbf{f}$ denote the desired and actual sensed forces, respectively, whereas $\mathbf{x}_e$ is the desired position input, and $\mathbf{K}_e$ is the environment stiffness. The end effector was controlled using its position controller based on the external forces that are sensed through its force sensor located at the wrist. This approach is simple and shows a high rejection of disturbances in the actuator system.

An outer interaction control loop was suggested that computes the reference end-effector trajectory for the robot's inner task space, and joint-space servo controllers ensure the motion control. The controller scheme is shown in Fig. 5.2b. The work was extended for impedance and parallel force/position controls that regulate both contact force and moment to a desired value assigned along the constrained directions. An open version of COMAU C3G 9000 control unit with Smart-3S robot was used to implement the modular, multi-layer, and real-time control structure. However, such an open controller architectures are rare in commercially available industrial robots.

Khatib et al. [5] presented a concept of *torque-to-position transformer* that allowed joint-torque control-based force controller schemes to be implemented as a software unit to any conventional joint-position-controlled industrial robot. With the knowledge of a joint-position servo controller and the closed-loop frequency response each joint-torque commands are converted to instantaneous joint displacements. This method was demonstrated on a Honda ASIMO robot arm.

All force control methods are required to have a comparable degree of passive compliance in order to yield a comparable execution speed and disturbance rejection [6]. A hybrid compliance with control switching for a humanoid shoulder mechanism was proposed in [7], which used an active compliance for low frequency domain and passive compliance for high frequency domain applications. A set of compliant actuators with passive adjustable compliance was discussed by [8], where the use of such actuator was suggested for safe human–robot interactions. Although such systems may be desired for intrinsically safe robots involving human accessible workspaces but are not suitable for precise positioning tasks like pick-and-place operations. Moreover, these systems do not allow passive compliance to be adjusted while the robot is moving. In [9], the authors used a KUKA KR5 Six robot with the same closed control architecture and Robot Sensory Interface (RSI) as in this chapter. However, they do not use a force sensor for active compliance, and still collision was detected by monitoring motor currents. A compliant-like behavior was realized by an online user-defined modifications of the position reference for the original KUKA controller. Moreover, gravity forces acting on the links were estimated and removed for a better sensitivity to collisions. A physical human–robot interaction using dual industrial robots was discussed by [10] where a sensorless estimation of external forces and saturation of joint control torques was used to keep the effective applied force to a safe level. However, the authors did not addressed any active force control at the robot's end effector. The peg-in-hole with Gaussian mixture model using a redundant manipulator having torque sensors at its joint is reported in [11].

This chapter details on parallel active–passive force control scheme which is suitable for compliant tasks like cooperative manipulation, surface finishing, and assembly operations. Here, the word "active" is implied as the force control based on the feedback from the F/T sensor located at the end effector, whereas "passive" is understood as a control scheme without any measurement of force or torque from sensors rather was induced by limiting the motor current. Moreover, the passive compliance reported in this chapter is different from a physical spring as it gets activated only when the robot is powered on. This makes the system safe against any pinching or trapping hazard [12] and reduces the hazards of impact on collision due to actuator forces, along with precise force–position control tasks it can perform. The system was implemented on a standard floor-mounted KUKA KR5 Arc industrial robot without any modifications to the existing joints or links of the robot. Recently, peg-in-hole robot assembly system based on Gauss mixture model was discussed in [11] with the motion planning with compliant control in [13].

5.1.2 Assembly

In order to raise the level of a typical industrial assembly situation, the work adopted here is *peg-in-tube*. Peg-in-tube forms the superset of industrial assembly task where the possibility of finding the target hole is obscured. This is because there exists a surrounding pocket which may be mistaken with the actual hole where the part is to be inserted. The existing well-researched peg-in-hole task forms a special case of the peg-in-tube task under study where the outer radius of the tube is infinity. Many such cases exist in the industry which may be pronounced as *peg-in-tube* task. For example, inserting a piston into the cylinder liner, inserting a plunger into a cylindrical fuel pump barrel, putting cylindrical fuel pellets into a tube in a nuclear power plant, fitting a shaft into a circular bush, etc. Most of the existing peg-in-hole strategies are inappropriate to search the hole location of a tube in its current form. Some of them can be modified beneficially to be used for peg-in-tube task. There are primarily two different approaches to the hole search problem. The first one uses the robot's end-effector position, whereas the second one uses a force/torque sensor fitted at the robot's end effector.

Kim et al. [14] precisely estimated the shape and location of the peg with respect to the hole using a six-axis force/torque sensor data. Relying entirely on force sensor data is not worthwhile in a typical industrial environment where the force sensor data is proven to have a very high noise [15]. These noise may be attributed to the mechanical vibration of the work floor, electrical noise, electromagnetic radiations, etc. For any two surfaces in contact and in relative motion, the measured contact force will depend largely on their surface finish and the force controller gains. Moreover, if the surface stiffness is high, a small relative displacement due to vibration will cause a high value of noise in the sensed force. The same applies to any force/moment-based method for hole detection, which will be more evident through an experiment described later in this chapter.

A blind search strategy for localization by generating a depth-map of the tilted peg center from the hole surface, and applying a particle filter to locate the hole was proposed by [16, 17], and many others in the past. This method has the potential to be used for "peg-in-tube" case as well, but the depth-map is required to be regenerated if the dimensions of the peg or the hole change. Generating a new map is time-consuming if it is prepared online using the robot holding the peg and plunging it over the hole surface by multiple number of times. Creating depth-map becomes complex if it is generated offline using simulations based on any analytical method. Due to the presence of pockets around the actual hole, finding a unique hole location on a tube by plunging the peg for a lesser number of times is ruled out, as this will give rise to multiple solutions. Moreover, plunging the peg for a larger number of times will increase the insertion time. This chapter would discuss an intuitive software framework developed in-house which can be used for quick depth-map generation. However, depth-maps may not be realized for actual localization if the environmental constraints like system compliance, surface finish, vibration, etc. are not taken into account for creating the depth-map along with the accurate calibration of the robot's tool frame that holds the peg.

A strategy for high-precision assembly in a semi-structured environment based on vision and force control was demonstrated by [18]. Detecting the pellet pose using machine vision suffers from calibration errors, blurred view for objects that are close by, improper illumination, noise due to the background environment, etc. Also, the time required to assemble with spiral search used by [18] depends on the initial offset of the peg from the peg center. A set of environment-independent search strategies was proposed by [17], where a neural-network-based strategy based on moments and descent of the peg into the hole was generalized for tilted peg. Such optimization techniques are suited well for offline simulations but it may prove to be non-realistic for a real-time insertion. Moreover, the hybrid force–position control [19] as suggested by [17] cannot be applied to a position-controlled industrial robot. Nemec et al. [20] demonstrated a high-speed assembly of peg and chamfered holes using learning on a joint-torque-controlled robots. Such robots, however, are rarely used in the industry where the robots are required to perform precision tasks with heavy loads. Therefore, a blind search technique is needed that can efficiently eliminate the possibility of peg being dropped outside the tube, provided it is brought sufficiently near to the hole using existing automation, to say, using a vision system.

This chapter proposes a hole search method for typical position-controlled industrial robot, which measures the depth of the tilted peg center from the top surface of the tube while the peg rotates about the axis perpendicular to the tube's top surface by one full revolution. The profile of this depth was used to find the direction of the hole. Note here that the positioning repeatability of an industrial robot is typically in the range of 100 μ m. For example, the KUKA KR5 Arc robot used in this work has ±0.10 mm, i.e., 100 μ m positioning repeatability, whereas most of the assembly tolerance is in the order tens of microns [14]. Position and orientation errors can cause jamming and wedging during insertion. An active force control, with passive joint compliance based on current limiting, is expected to solve this problem by accommodating any minute alignment error during insertion. However, the current work assumes the peg is aligned with the tube using any suitable method like [21]. To simplify the problems presented above, this chapter uncovers the geometrical aspects and the practical considerations that are required to be understood before dealing with chamferless cylindrical peg-in-tube task. This chapter also proposes a novel algorithm to tackle "peg-in-tube" task. Experimental results presented here show the viability of the proposed algorithm.

Based on the literature discussed following objective are needed to be fulfilled:

- A force controller scheme for existing joint-position-controlled robots.
- Make the robot joint compliant to ensure safety against any unpredicted link collision which will be useful during the assembly task.
- Implement an active wrist/end-effector force control for precise assembly tasks.
- A blind hole search technique for peg-in-tube task.

The following sub-sections discuss these implementations.

5.2 Passive Force Control and Joint Compliance

In this section, the approaches for the current control of industrial manipulator are discussed. Having a limited current at the joints enables it to become compliant when enacted by any uncertain load. Thus, the joints are henceforth called as *passive compliant joint* as it behaves like the one fitted with a preloaded torsional springs. A KUKA KR-5 Arc industrial robot is used to demonstrate this control algorithm which allows limiting of currents flowing through joint actuators using special commands. The run-time torque during operations will be due to static and dynamic forces. As most of the industrial robot-to-environment interaction tasks are slow, the following sub-sections approach to identify the minimum current required to attain the torque required to maintain the robot in its various commanded poses while in slow moving condition.

5.2.1 Current Sensing Due to Gravity

The controller introduces a compliant behavior of the robot by limiting the current supplied to the motors and thereby limiting the maximum torque it can deliver. This makes the robot's joints behave like a torsional spring when it encounters an obstruction due to interaction with any unintended object in the robot's workspace. The torque delivered by the joint motors is just sufficient to enable the robot links to move with the desired trajectory or to be in a static pose. So, it becomes necessary for the controller to calculate current required by the motors to counterbalance the weight of the system against the gravity forces and to maintain a given trajectory. The first part of the controller design commence with identification of the mass moments of the robot links. A backward recursive technique for mass moment calculation as in [22] was used to estimate the mass moments of the links. With this the joint torque due to gravity can be estimated. Later, the relation between the joint actuator's current and the torque corresponding to a trajectory or pose of the robot inside the robot's workspace was estimated. This enabled to estimate the necessary actuator current which is required to compensate against the gravity forces. The sequence of identification was thus comprised of identification of joint angle-to-torque relationship, and finally joint torque-to-current relationship. This was done in order to obtain a closed-form analytical relation between the joint angle and its corresponding current. The identified model now allows easy interpretation of the joint-current characteristics corresponding to any joint angle. As most of the force-controlled task are performed slow, it is reasonable to assume any torque or current due to dynamic forces as negligible.

5.2.2 *Formulation*

The joint-torque and angular position values were obtained from KUKA Robot Sensor Interface (RSI) through Ethernet as explained in [23, 24]. Although the robot is not equipped with explicit joint-torque sensor, the RSI provides the joint torques, which is based on the internal unpublished model parameters of the robot. As the robot was configured as a floor-mounted robot, the torque obtained by the RSI was that for the floor-mounted robot. To estimate the variation of the torques with respect to the angles, the joints were provided with periodic motions. The robot was tested with motion in the sagittal plane for pick-and-place operation may be after it has rotated about the first axis without changing the orientation. The above motion restriction may limit some application domain where the orientation needs to be changed while placing it on a desired location. However, if the objects are cylindrical and pick/place operations are done on a horizontal plane then the assumptions of the robot's motion do not cause any problem. Hence, no motion was provided to joints 4 and 6, as shown in Fig. 5.3. The joint axis 1 is parallel to the direction of acceleration due to gravity $\mathbf{g}$ and thus moving joint 1 does not influence the joint torques due to gravity. Motions were given to joints 2, 3, and 5 only as they are always orthogonal to the gravity vector.

Figure 5.4 shows two generic links, say, $(i-1)$th and ith link written as $\#(i-1)$ and $\#i$ of the robot, respectively. According to the Denavit–Hartenberg (DH) convention [25], $\#i$ is between joints i and $(i+1)$, whereas $\#(i-1)$ connects the joints $(i-1)$ and i. Let $\mathbf{o}_i$ and $\mathbf{o}_{i-1}$ be the position vectors of the origins of $\#i$ and

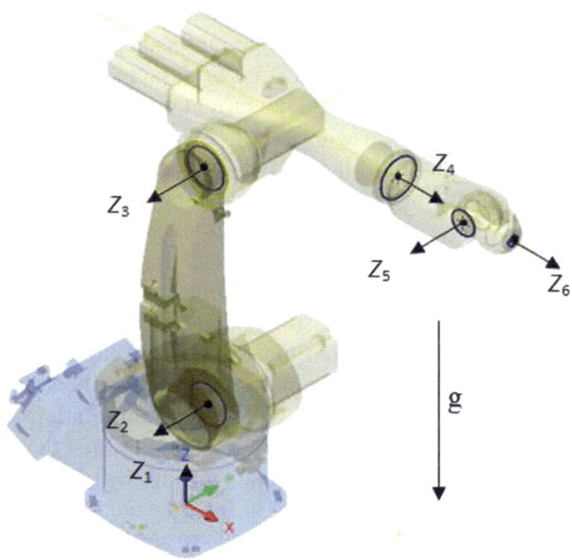

Fig. 5.3 Joint axis of KUKA KR5 Arc robot

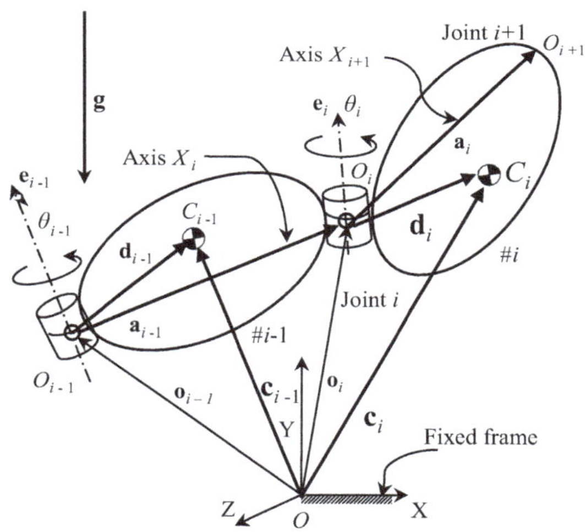

Fig. 5.4 Coupled links of a robot

#$(i-1)$, i.e., O_i and O_{i-1}, respectively, where the joints are coupling the links. Let $\mathbf{a}_{i-1}$ denote the position vector of the origin of #i with respect to #$(i-1)$ and $\mathbf{d}_i$ be the vector joining the origin O_i to the center of gravity of the #i, i.e., C_i of Fig. 5.4. Note that the actuator torque for the ith revolute joint, τ_i, can be obtained by projecting the driving moments onto their corresponding joint axes. For static moment due to gravity only, i.e., $[\mathbf{n}_{i,i+1}]_i$, it is given by

$$\tau_i = [\mathbf{e}_i]_i^T [\mathbf{n}_{i,i+1}]_i \text{ : for revolute joint} \tag{5.1}$$

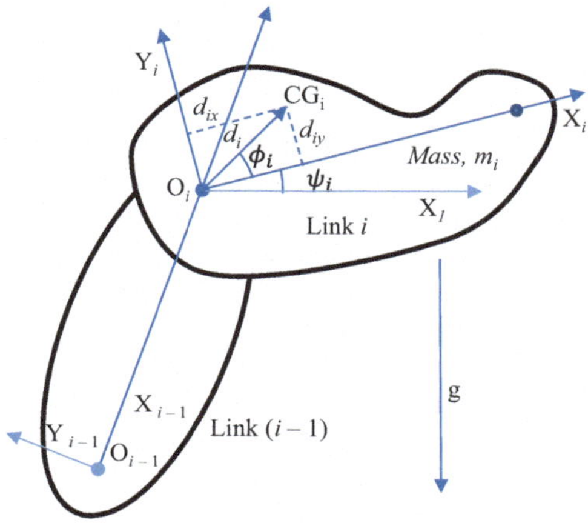

Fig. 5.5 Location of CG on a link

where $[\mathbf{n}_{i,i+1}]_i$ is the moment on #i by #$(i+1)$ which is represented in Frame i, whereas $[\mathbf{e}_i]_i$ is the unit vector parallel to the axis of rotation of the ith joint represented in the ith frame, i.e., $[\mathbf{e}_i]_i \equiv [0\ 0\ 1]^T$ [25]. Moreover,

$$[\mathbf{n}_{i,i+1}]_i = m_i \left[\mathbf{d}_i \times \mathbf{g}\right]_i \tag{5.2}$$

Expanding the vector representation in Frame i, (5.1) can be rewritten as

$$\tau_i = m_i d_i \mathrm{g} \sin(\psi_i \pm \phi_i) \tag{5.3}$$

where ϕ_i is the angle subtended by the projection of $\mathbf{d}_i$ on the plane perpendicular to $\mathbf{e}_i$ from the link axis X_{i+1}, and ψ_i is the swept angle of link i with respect to the horizontal as shown in Fig. 5.5. The expression of τ_i above can be recursively calculated from the outer link to the inner link as

$$\tau_{i-1} = \tau_i + m_i \left[\mathbf{d}_i \times \mathbf{g}\right]_{i-1}^T \left[\mathbf{e}_{i-1}\right]_{i-1} \tag{5.4}$$

From (5.4) it may be stated that torque at any joint due to gravity is a function of joint angles of all the links moving along the gravitational plane. Hence, $\tau_i = f(\theta_2, \theta_3, \theta_5),\ i = 2, 3, 5$, as long as the robot lies in the sagittal plane.

5.2.3 Identification of Mass Moments

It was observed that the variation of the torques is sinusoidal in nature as the joints were moved. This is in agreement with the analytical expression of torque due to gravity Eq. (5.3). Fourier approximation as in [26] was used to best fit the curves obtained by changing the angle ψ in Eq. (5.3). This is expressed as

$$\tau_i = a_0 + \sum_{n=1}^{\infty} [a_n \sin(n\psi) + b_n \cos(n\psi)] \tag{5.5}$$

where τ_i is the torque at the ith joint, moreover a_0, a_n, and b_n are the coefficients of the series function that may also be related to the constant terms on expanding Eq. (5.3), n is an integer, and ψ is the angle swept by the links. The first-order approximation was taken by considering the value of $n = 1$. Using MATLAB's function `fit(x,y,fitType)`, where `x` is the independent variable (in this case θ's), `y` is the dependent variable (here τ 's), and `fitType` is the order approximation (here it is one). The function returns the value of the Fourier coefficients.

The process of identification started by giving motion to the distal link first. As the link #6 is symmetrical about its joint axis, moving it would cause no torque changes. The motion was given to the joint 5 which moves links #5 and #6 together, and the joint torques were recorded. Since the torque at joint 5, τ_5 actually changes by changing any of the joint angles 5, 3, or 2, it may be expressed as a function of 5th, 3rd, and 2nd joint angles, i.e., $\tau_5 \equiv f(\theta_5, \theta_3, \theta_2)$. The data obtained for τ_5 and the combined swept angle $\psi = \theta_5 + \theta_3 + \theta_2$ was fitted using Fourier fit function to obtain

$$\begin{aligned} \tau_5 &= -0.052 - 11.47\cos(\psi) + 4.9\sin(\psi) \\ \tau_5 &\approx -11.47\cos(\psi) + 4.9\sin(\psi) \end{aligned} \tag{5.6}$$

Since the magnitude of the coefficient a_0 $(= -0.052)$ in the expression for τ_5 is very small, which may be attributed to the constant torque required mainly to overcome the friction in the joints. It was neglected for further use. Comparing (5.3) with (5.6), it was observed that the coefficients of sine and cosine can be physically represented as the components of the mass moment of the links. Table 5.1 lists the identified mass moments, where d_{ix} and d_{iy} are the components of the vector d_i joining the origin O_i and the center of gravity CG_i as shown in Fig. 5.5. Similarly, the torques in joint 3 and 2 denoted with τ_3 and τ_2 were recorded and expressed using (5.3), (5.4), and (5.5) as

$$\begin{aligned} \tau_3 &= -126.28\cos(\theta_3 + \theta_2 - 44.85) + \tau_5 \\ \tau_2 &= -375.70\cos(\theta_2) + \tau_3 \end{aligned} \tag{5.7}$$

The robot was moved with cyclic trajectory for the initial and final values mentioned in Table 5.2. The empirical torque relations (5.6) and (5.7) were used to estimate the

Table 5.1 Mass moments of the links

Link i	$m_i d_{ix} g (Nm)$	Mass moments $m_i d_{ix}$	$m_i d_{iy} g$ (Nm)	Mass moments $m_i d_{iy}$
5	11.47	1.169	90	0.499
3	89.49	8.612	88.86	9.058
2	375.70	38.297	0	0

Table 5.2 Joint angles to test the empirical torque values

	Joint 1	Joint 2	Joint 3	Joint 4	Joint 5	Joint 6
Initial	90	−45	30	0	−90	0
Final	90	−90	50	0	90	0

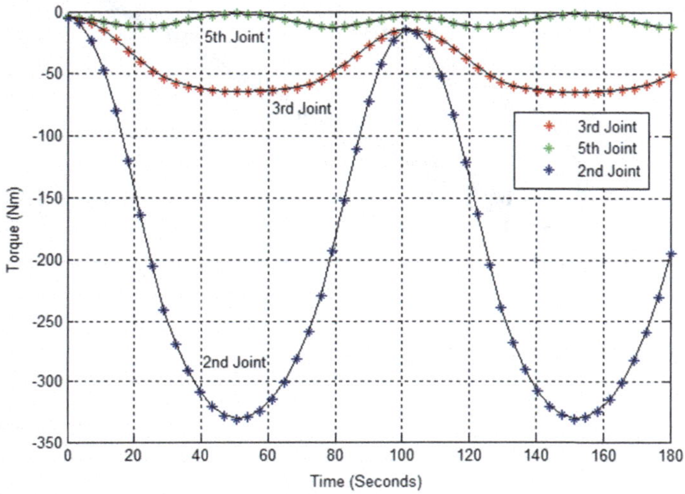

Fig. 5.6 Variation of estimated and actual joint torques

torques which were found to be in agreement with the recorded torques, as shown in Fig. 5.6.

5.2.4 *Estimating current limits*

The torque and current variation with the joint motion was recorded through RSI. Figure 5.7 shows the variation of current corresponding to the torque for an angle varying cyclically from 0 to 90° for joint 3. It may be observed that the current varies proportional to the torque at the joint. However, it is different in clockwise from anticlockwise motion of the joint. This behavior of the current to torque depicts a hysteresis loop and was approximated as

$$I_i = \frac{\tau_i}{k_m} + \alpha_i + sgn(\dot{\theta}_i)\beta_i \tag{5.8}$$

where k_m (Nm/A) is the motor torque constant, $sgn()$ function returns the sign of ith joint velocity $\dot{\theta}_i$, and α_i and β_i are empirical constants obtained by linear fit. The constant β_i signifies the offset from the mean current when the torque is zero. It is required to drive the motor against the joint friction. Table 5.3 lists the characteristic

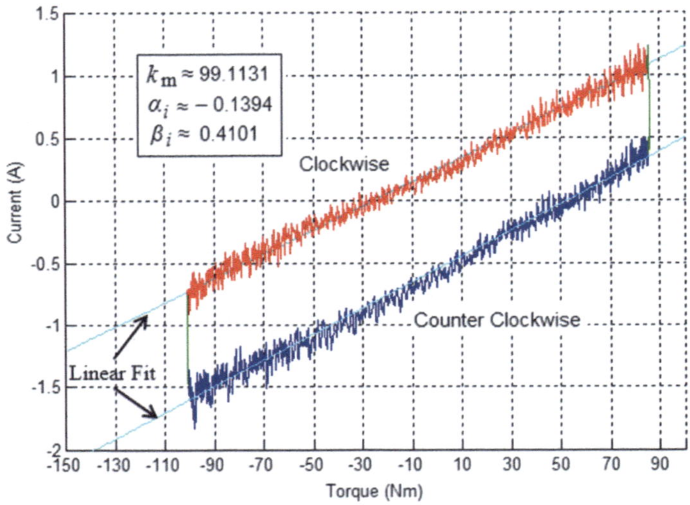

Fig. 5.7 Variation of current with torque for Joint 3

Table 5.3 Empirical constants for motors

Motor i	k_m (Nm/A)	α_i (A)	β_i (A)
5	93.153	0.036	0.249
3	99.113	−0.139	0.410
2	220.220	0.118	0.338

parameters of motors 5, 3, and 2. The empirical relations (5.6), (5.7), and (5.8) then enable one to estimate the joint currents required to maintain the robot's pose in the sagittal plane. As the joints 1 and 6 require constant torque or current to drive the joints at a particular speed (joint 1 being parallel to **g** and 6 being symmetrical along its axis), a fixed current was allowed to flow through these joint motors. In order to make the joints compliant against any additional torques due to external forces on the link, the current was limited to 10% in excess of the minimum current required to maintain the robot's pose against gravity while moving freely. The robot may be left without any current limits when the end effector establishes a contact, and active force control comes into action.

5.3 Active Force Control

The overall control scheme proposed in this chapter for active–passive force control is depicted in Fig. 5.8, where $G(s)$ is the force control law in Laplace domain denoted by character s, $H(s)$ is the transfer function for the closed-loop servo controller of

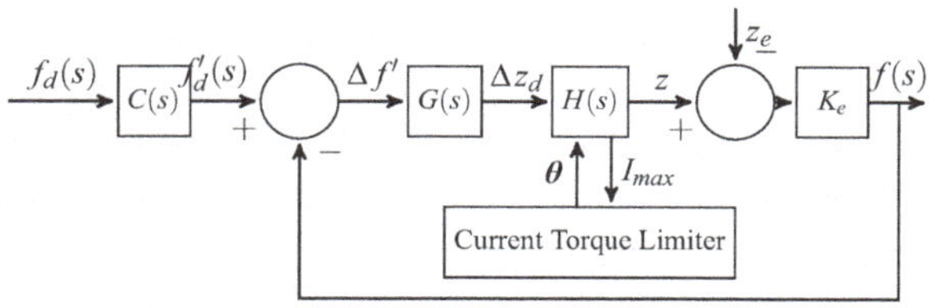

Fig. 5.8 Active–passive force control scheme [28]

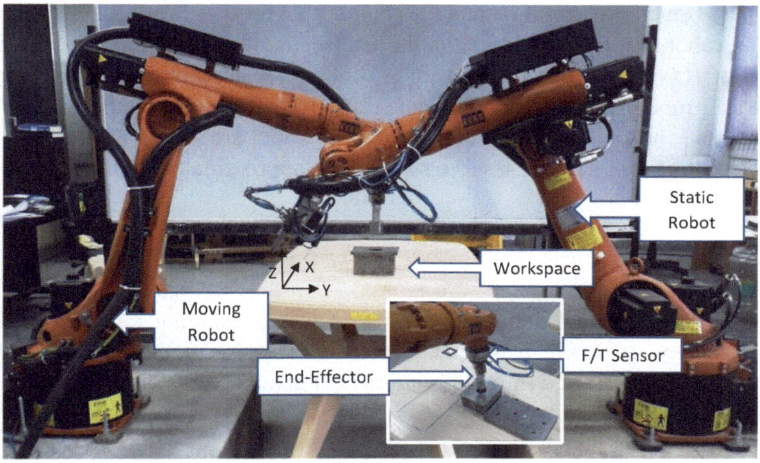

Fig. 5.9 Experimental setup for robots

the robot, and k_e is the contact stiffness. The force controller was purely proportional in nature as suggested by [27] for a low payload capacity six-axis industrial robot manufactured by KUKA. In this work, a similar capacity (5 Kg payload) industrial robot KUKA KR5 Arc equipped with a six-component F/T sensor manufactured by SCHUNK at its end effector was used. Figure 5.9 shows the experimental setup that shows two robots, their shared workspace, and the F/T sensor. The force controller calculates the end-effector displacement Δz_d corresponding to the force error $\Delta f'$. The position control is done internally by the robot controller, i.e., $H(s)$ in Fig. 5.8. This generates the end-effector position denoted by z. Also, z_e denotes the position of the contact environment. The contact stiffness k_e includes stiffness caused by the F/T sensor, workbench, and the entire end effector with the gripper system. The following sub-sections discuss the identification of block transfer functions $H(s)$ and k_e, which are prerequisites for efficiently defining the force controller gains and finally the compensator $C(s)$.

5.3.1 *Position Servo Controller, $H(s)$*

The robot was fed with a position correction Δz_d through RSI Ethernet. The position servo controller $H(s)$ creates the desired set position z_d using the actual end-effector position z_a available with the controller as

$$z_d = z_a + \Delta z_d \tag{5.9}$$

The value of desired set position and the actual robot position was accessed using the RSI. The robot was fed with a step input displacement of 24 mm from the existing position of 1006 mm along the vertical Z-direction. The maximum velocity of the robot was 2 m/s to attain the desired position of 1030 mm. The desired step input trajectory and the actual end-effector position of robot are shown in Fig. 5.10. This behavior of the robot's position servo system was approximated with a second-order system, as in [4], i.e.,

$$H(s) = \frac{k\omega_n^2}{s^2 + 2\xi\omega_n s + \omega_n^2} \tag{5.10}$$

where ω_n is the damped natural frequency, ξ is the damping ratio, and k is the static sensitivity. Using Fig. 5.10, maximum overshoot was measured as 0.7153 mm, whereas ξ can be extracted through relation [29] given below:

$$\text{Overshoot (OS)} = (z_{\text{initial}} - z_{\text{final}})e^{-\frac{\xi\pi}{\sqrt{1-\xi^2}}} \tag{5.11}$$

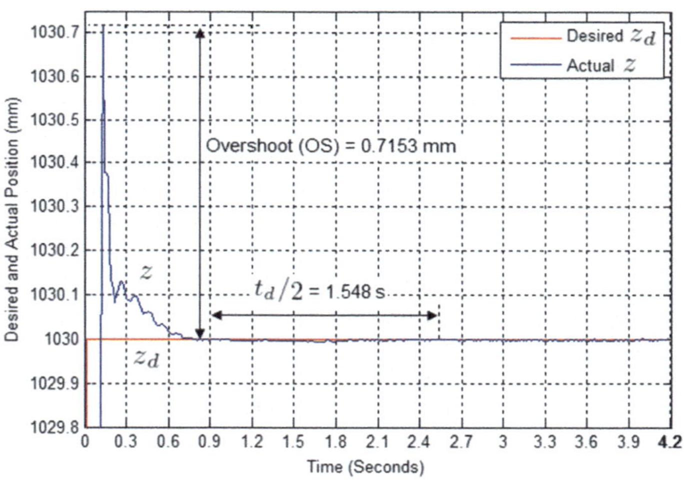

Fig. 5.10 End-effector position of robot for a step input

From (5.11), the damping ratio was extracted as $\xi = 0.7492$. Similarly, the period of oscillation t_d was measured as 3.096s from the zoomed-in graph of Fig. 5.10, which is nothing but the time between the two successive peaks of the response plot. The damped natural frequency ω_n can thus be obtained using

$$\omega_d = \frac{2\pi}{t_d} = \omega_n\sqrt{1-\xi^2} \tag{5.12}$$

which gives $\omega_n = 3.064$ rad/s. The static sensitivity k was then calculated as

$$k = \frac{\Delta z}{\Delta z_d} \approx 1.0 \tag{5.13}$$

The above calculations allow one to completely define the robot's position transfer function $H(s)$. A time lag for communication was observed as 2 clock cycles, i.e., 24 ms for KUKA KRC2 controller with RSI. The proposed transfer function expresses $H(s)$ better than an exponential lag function as suggested by [27]. This was possible as overshoot and small oscillations were normally observable for any step inputs tested in the current work. However, for steady-state condition, i.e, where $s \to 0$, for both systems would result in a unity gain, i.e., $H(s) = 1$. A step response method is proposed for identification of $H(s)$ as the robot will be encountering a similar condition for a contact situation.

5.3.2 Estimation of Contact Stiffness, k_e

Assuming low velocity during force control task, a simple spring behavior is proposed by various researchers as a contact model of the environment interactions. This may be expressed as follows:

$$f = k_e(z - z_e),\ \forall z > z_e;\ 0\,\forall z \leq z_e \tag{5.14}$$

As the end effector was mounted with a F/T sensor, it was lowered into the workspace along negative Z-direction of Fig. 5.9 and the contact was established. The force readings corresponding to the displacements of the end effector were obtained, from where the combined stiffness k_e was estimated by taking mean of 10 readings as 23.38 N/mm.

5.3.3 Force Control Law and $G(s)$

The scheme comprises external force control loop as in [4]. This is indicated in Fig. 5.8. The response of the one-dimensional force control loop with two inputs, i.e., f_d' or z_e, is given by

$$f(s) = \frac{k_e G(s) H(s)}{1 + k_e G(s) H(s)} f_d'(s) - \frac{k_e}{1 + k_e G(s) H(s)} z_e(s) \tag{5.15}$$

To analyze the steady-state error, it was assumed that the environment is fixed, i.e., the workbench of Fig. 5.9 and the robot has already established the contact, i.e., $z_e = 0$. The force controller is purely a proportional one as in [27] with gain k_p which implies $G(s) = k_p$. The robot transfer functions $H(s)$ and k_e were estimated as discussed in Sect. 5.3.1. Substituting these in (5.15) the steady-state (SS) error Δf_{SS} may be written as

$$\Delta f_{SS} = f_d' - f = \frac{f_d'}{1 + k_p k_e} \tag{5.16}$$

where f_d' is the desired input force to the standard external force control loop shown in Fig. 5.8. It may be observed in (5.16) that the steady-state error depends on the desired force itself. So, in order to eliminate the steady-state (SS) error, a SS error compensator block was proposed for the external force control scheme shown in Fig. 5.8. This allows one to set the desired force f_d so that the compensator block actually feeds in a force f_d' to the external force control loop and the controller achieves the actual force f_d desired. The transfer function for the compensator block $C(s)$ can be back calculated using $f = f_d$ in (5.16) as

$$f_d' = C(s) f_d \tag{5.17}$$

$$\text{where, } C(s) = \frac{1 + k_p k_e}{k_p k_e} \tag{5.18}$$

This completes all the block calculations for the proposed active–passive parallel controller scheme shown in Fig. 5.8. The controller was implemented using KUKA commands as briefed next.

5.3.4 Implementation in KUKA Controller

The controller was programmed in Microsoft C#.Net on a standard PC, whereas KUKA was communicated through TCP/IP using ST_ETHERNET command of KUKA.RSI 2.3 installed on KUKA KRC 2 robot controller. Position corrections corresponding to active forces were sent to the robot using ST_SKIPSENS which produces the actual Cartesian trajectory fed by the linked ST_PATHCORR object. In order to set or update the current limits, the RSI was programmed to use ST_BREAKMOVE to come out of the continuous ST_SKIPSENS using a flag sent by the PC. The KRL command $TORQUE_AXIS was used to select the axis where current limits are to be used. The values for current limits were set using

$CURR_RED command of KRL. Once the values were set, a PTP $POS_ACT command was used and the controller was again made to enter into continuous ST_SKIPSENS path correction mode.

5.4 Results and Discussion on Force Control Strategy

The controller was tested with set force ranging from $10N$ to $60N$. Figure 5.11 shows the response of the controller with $k_P = 0.005$ mm/N and the set force of $20N$. The robot was made to move with pure position control at maximum velocity of 0.2 m/s in free space while the forces were being monitored, i.e., a *guarded move* approach. A maximum overshoot of 4.75% was observed when the end effector established contact with the workspace at time 2.09s. The proposed controller also kept monitoring the joint angles and estimates the required current for each actuator. The system was then allowed only 10% in excess of the required current for that pose. This made the joints compliant against external forces during the free movement and during collision without affecting the position control performance while moving in free space. Figure 5.12a shows the variation of actuator current when a robot maintaining a position and encounters a link collision with another robot moving in its workspace. It may be observed that the actuators at axis 1 and 3 attain its saturation current limits of $0.48A$ and $-2.28A$, respectively, while it flexes. Joint actuators at axis 2 and 4 show significant rise in current. Similarly, Fig. 5.12b shows the variation of current when a moving robot is obstructed by a human operator. Such could be the case when a human operator gets trapped within the workspace of the robot. It

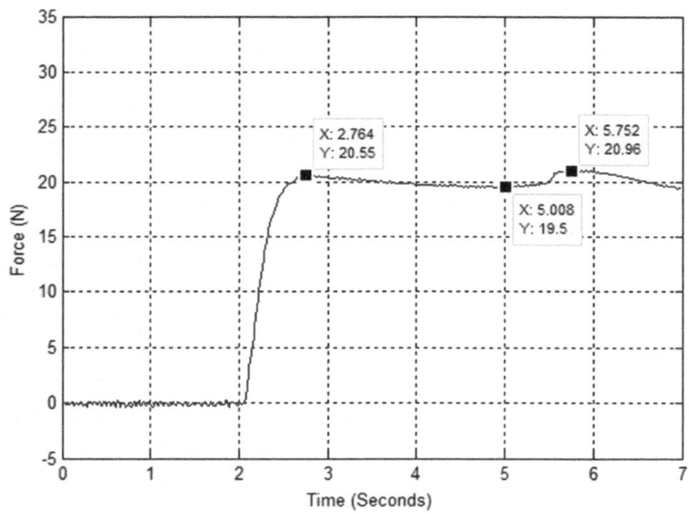

Fig. 5.11 Step response with a set force of $20N$

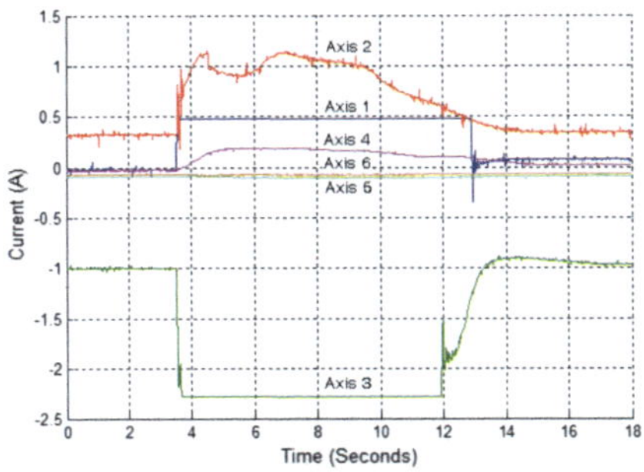

(a) Static robot hit by a moving robot

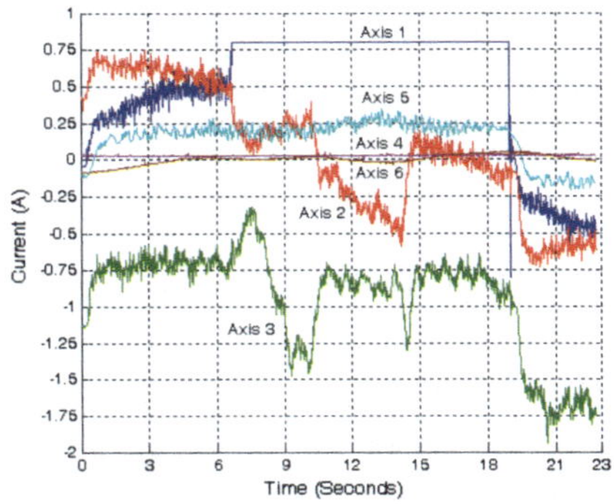

(b) Moving robot obstructed by an operator

Fig. 5.12 Robot interactions with its environment

may be observed that the current drawn by axis 1 reaches the maximum limit of $0.7998A$ and stalls, while other actuators are disturbed from the actual trajectories. The experimental setup for the two robots sharing a common workspace is shown in Fig. 5.9. An external force on individual links was applied by a human operator and saturation current was observed at joints which made the robot passively compliant. The parallel active–passive force controller scheme discussed here was implemented for the peg-in-tube task discussed next.

5.5 Peg-in-Tube Task and Its Geometrical Analysis

Peg-in-tube assembly stands ahead of a more common benchmark task for industrial assembly, i.e., "peg-in-hole". The robot can easily be deceived to detect the actual hole while performing a "peg-in-tube" task as the tube has a surrounding pocket that cannot support the peg.

The proposed search procedure involves rotating a tilted peg about the axis which is perpendicular to the tube's top surface and passing through peg's bottom center. The peg maintains a continuous contact with the tube during the search phase. Thus, the peg moves in a cone with a half-angle of the vertex equal to the amount of the tilt angle θ, as shown in Fig. 5.13.

This section will investigate the geometrical aspects of such contact and will extract the required peg center depth c_z from the tube's top surface, as shown in Fig. ??. This depth information was utilized later to find the hole direction. A tilted peg whose center lies outside the hole center with an offset $(c_x,\ c_y,\ c_z)$ can make contact with the tube in four different states. They are shown in Fig. 5.14a–d. There are two other states in which the peg can make contact with the hole. Firstly, when the curved surface of the peg comes in contact with the tube's inner rim, as shown in

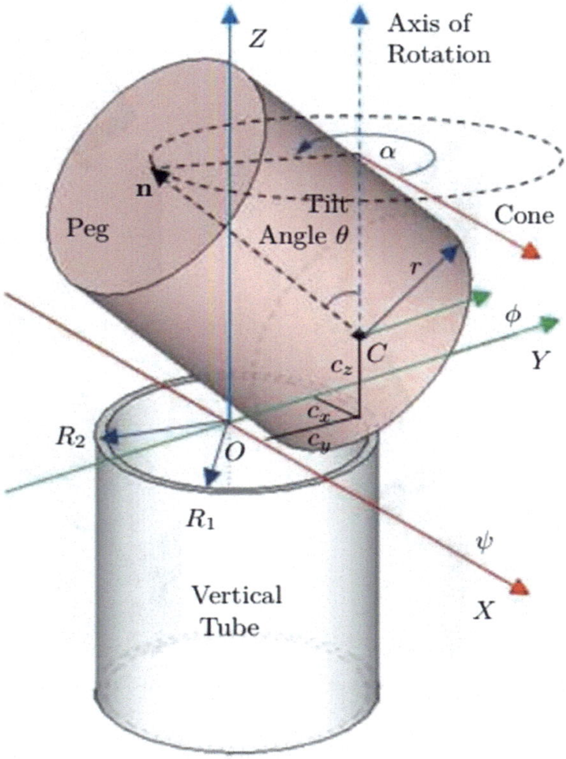

Fig. 5.13 A tilted peg in contact with the tube

Fig. 5.14e, and secondly when the peg's bottom cap comes in contact with the tube's inner curved surface, as shown in Fig. 5.14f. Latter two cases do not arise during the search procedure proposed in this chapter and hence it is not analyzed.

5.5.1 Parametric Modeling of the Peg and Tube

In this section, we define the peg and the tube surfaces, and their edges before we proceed to establish the conditions of peg lying inside, outside, or on the tube. The bottom of the peg was defined by a parametric equation of a three-dimensional circle $\mathbf{p}$. This is given by

$$\mathbf{p} = \mathbf{c} + \mathbf{u}r\cos\beta + \mathbf{v}r\sin\beta \tag{5.19}$$

where $\mathbf{u}$ and $\mathbf{v}$ are the unit vectors that lie on the peg's bottom face with origin at the face center and perpendicular to the vector $\mathbf{n} \equiv \mathbf{u} \times \mathbf{v}$, as shown in Fig. 5.15a. The radius of the peg is r and $\mathbf{c}$ is the position vector of the peg's bottom center. The

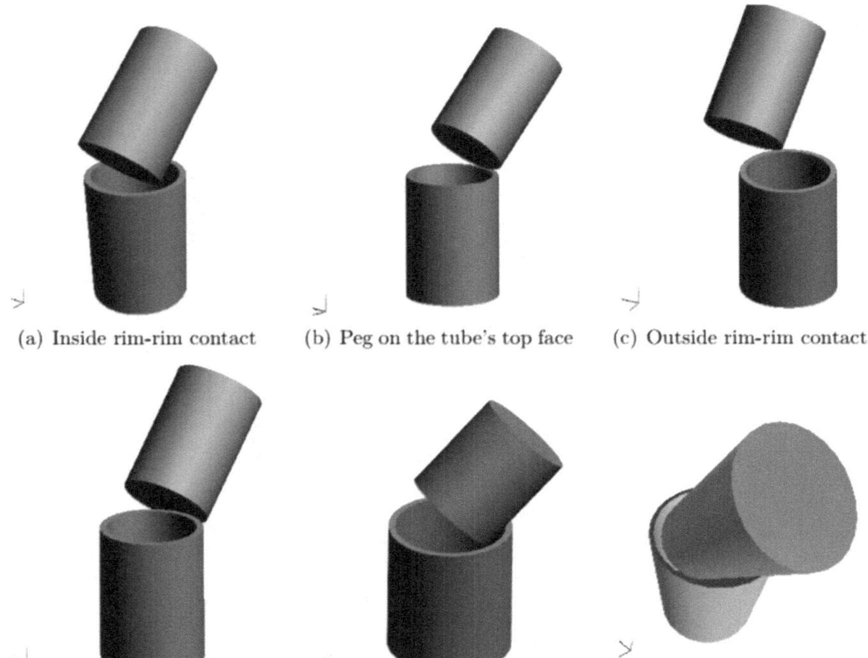

(a) Inside rim-rim contact (b) Peg on the tube's top face (c) Outside rim-rim contact (d) Bottom cap of the peg on the tube (e) Inside rim-body contact (f) Inside cap-body contact

Fig. 5.14 States of contact for Peg-in-Tube

parametric angle β is measured about $\mathbf{n}$ with respect to $\mathbf{u}$. For each $0 \leq \beta \leq 2\pi$, $P(x, y)$ represents the peg's bottom rim. Any tilt about X- and Y-axes was obtained by applying the transformation as

$$\mathbf{u}' = \mathbf{Q}_x\mathbf{Q}_y\mathbf{u} \text{ and } \mathbf{v}' = \mathbf{Q}_x\mathbf{Q}_y\mathbf{v} \tag{5.20}$$

where $\mathbf{Q}_x$ and $\mathbf{Q}_y$ are the 3×3 rotation matrices about X and Y axes, respectively. To start with $\mathbf{u} = [1\ 0\ 0]^T$ and $\mathbf{v} = [0\ 1\ 0]^T$ were taken, such that $\mathbf{n} = [0\ 0\ 1]^T$. Thereafter, the rotations were applied so as to obtain any given tilt angle. Now, for any point $P(x,\ y,\ z)$ having position vector $\mathbf{p}$ at the bottom circle of the peg, the condition $(\mathbf{p} - \mathbf{c}) \cdot \mathbf{n} = 0$ holds true as the vector $(\mathbf{p} - \mathbf{c})$ lies on the plane for which the normal $\mathbf{n} = [n_x\ n_y\ n_z]^T$ passes through $\mathbf{c} = [c_x\ c_y\ c_z]^T$. This condition is essentially the vector expression for representing the bottom-end of the peg. This may be written in scalar form as

$$n_x(x - c_x) + n_x(y - c_y) + n_z(z - c_z) = 0 \tag{5.21}$$

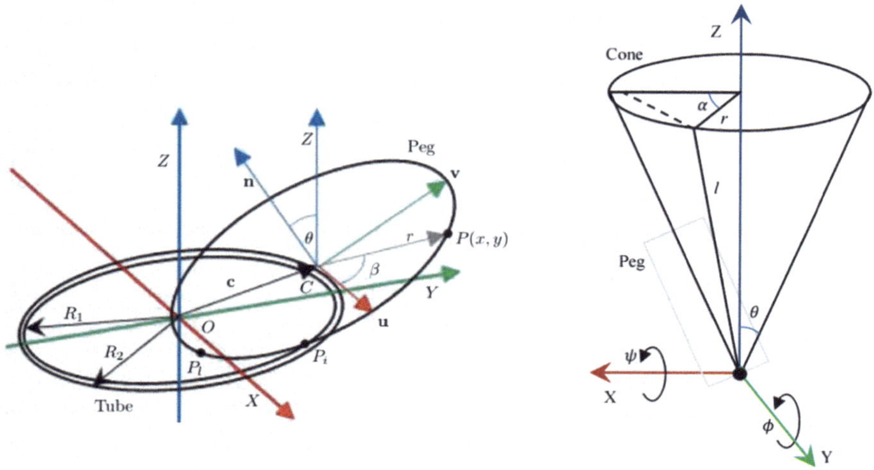

(a) Three dimensional circle

(b) Conical path of the peg during depth profile based search

Fig. 5.15 Three-dimensional circle and conical path

With this any contact with the peg's bottom surface or the edge can be geometrically explained. For that, the parametric equations for the tube hole are given next as

$$x = R_1 \cos(\xi) \text{ and } y = R_1 \sin(\xi) \tag{5.22}$$

where $0 \leq \xi \leq 2\pi$ and R_1 is the inner radius of the tube, as shown in Fig. 5.15a. Similarly, the outer edge of the tube with radius is R_2.

5.5.2 *The Rotating Tilted Peg*

Let us assume that the peg has a small tilt angle of θ and rotates about the vertical axis by an angle α, and l be the slant height of the cone, as shown in Fig. 5.15b. From Fig. 5.15b, it can be inferred that

$$\begin{aligned} r \sin\alpha &\approx l \sin\psi \\ r \cos\alpha &\approx l \sin\phi \end{aligned} \tag{5.23}$$

By substituting $r = l \sin\theta$ in Eq. (5.23) the tilt angle about global X- and Y-axes given by ψ and ϕ, respectively, may be expressed as

$$\psi = \arcsin(\sin\alpha \sin\theta) \text{ and } \phi = \arcsin(\cos\alpha \sin\theta) \tag{5.24}$$

where $0 \leq \alpha \leq 2\pi$. The projection of the peg's bottom cap forms a rotated ellipse and it constitutes an important aspect to identify the point of contact. The modified equation of the standard ellipse E to express the projected base of the peg may be expressed as

$$E = \frac{(\tilde{x})^2}{a^2} + \frac{(\tilde{y})^2}{b^2} - 1 \tag{5.25}$$

where $a = r$ and $b = r\cos\theta$ are the major and minor axes, respectively. Moreover, $\tilde{x}$ and $\tilde{y}$ are given by

$$\begin{bmatrix} \tilde{x} \\ \tilde{y} \end{bmatrix} = \begin{bmatrix} \cos\alpha & \sin\alpha \\ -\sin\alpha & \cos\alpha \end{bmatrix} \begin{bmatrix} x - c_x \\ y - c_y \end{bmatrix} \tag{5.26}$$

The state of the peg may now be completely defined by θ, α, and the offset $\mathbf{c} = [c_x\ c_y\ c_z]^T$ from the tube center which was taken as origin O here. The lowermost point of the peg P_l $(x_l,\ y_l,\ z_l)$ was given by the 3D circle equation, Eq. (5.19), at $\beta = \alpha$. As shown in Fig. 5.15a, the lowermost point is at $\beta = \alpha = -\pi/2$. The point of intersection of the projected peg's base and the tube P_i $(x_i,\ y_i,\ 0)$ was obtained by solving E of Eqs. (5.22) and (5.25). The potential point of contact is point P_i which is nearest to P_l. The possible location of P_i for a sample state of the peg is shown in Fig. 5.15a.

5.5.3 *Peg and Tube in Contact*

The cases discussed here completely define the rotation of the tilted peg about a vertical axis passing through the peg's bottom center, while the peg also maintains a constant contact with the tube during the rotation. The state of the peg for each case was examined individually and the depth information was extracted accordingly. They are explained below.

5.5.3.1 Peg Lies Inside the Tube

The projection of lowermost point of the peg P_l on the tube's top surface lies within the tube's inner circle, i.e., the hole, as shown in Fig. ??. This can be assured by checking, if the condition $x^2 + y^2 - R_1^2 < 0$ is true for the point P_l. The point of intersection of the projected ellipse E and tube's inner rim P_i that corresponds to the point of contact (the one which is nearer to P_l) satisfies the equation of the plane in Eq. (5.21). Thus the depth c_z may be calculated from Eq. (5.21) as

$$c_z = \frac{1}{n_z}[n_x(x_i - c_x) + n_y(y_i - c_y)] \tag{5.27}$$

5.5.3.2 Peg lies on the tube

The lowermost point lies between the inner and outer circular rims of the tube. Thus, P_l makes $x^2 + y^2 - R_1^2 \geq 0$ and $x^2 + y^2 - R_2^2 \leq 0$, and the depth c_z from Fig. 5.15a may be given by

$$c_z = r \sin\theta \tag{5.28}$$

This is illustrated in Fig. 5.14b.

5.5.3.3 Peg Lies Outside the Tube

The lowermost point lies outside the outer circular rim of the tube. Thus, P_l makes $x^2 + y^2 - R_2^2 > 0$. In this case, the peg may have a rim-rim contact case, as shown in Fig. 5.14a or rim-face contact case, as shown in Fig. 5.14d. In the case when the peg's cap comes in contact with the outer tube rim, the line joining the center of the tube O (Fig. 5.15a) to the point of contact P_t $(x_t,\ y_t,\ 0)$ becomes parallel to the projected normal of the cap's plane. This gives rise to the simultaneous conditions

$$\frac{x_t}{y_t} = \frac{n_x}{n_y} \text{ and } x_t^2 + y_t^2 = R_2^2 \tag{5.29}$$

Solving which we get the point P_t $(x_t,\ y_t,\ 0)$ as

$$\begin{aligned} x_t &= sign(n_x)\sqrt{\frac{R_2^2}{1 + (n_y/n_x)^2}} \\ y_t &= sign(n_y)\frac{n_y}{n_x}x_t \end{aligned} \tag{5.30}$$

For, $n_x = 0$, $x_t = 0$. Hence, the two sub-cases may be further investigated as

- Rim-rim contact: P_t lies outside the ellipse E and makes $E > 0$, as shown in Fig. 5.14c. Thus, P_i, i.e., the point of intersection of E and tube's outer circle lies on the peg's bottom rim and it satisfies the cap's plane equation, Eq. (5.21). Hence, from Eq. (5.21)

$$c_z = \frac{1}{n_z}[n_x(x_i - c_x) + n_y(y_i - c_y)] \tag{5.31}$$

- Rim-face contact: P_t lies inside the ellipse E which is the projected bottom cap of the peg shown in Fig. 5.14d. Thus, for P_t lying on the peg's cap plane, P_t makes $E < 0$. On substitution of $P_t = [x_t\ y_t\ 0]^T$ in Eq. (5.21), we get

$$c_z = \frac{1}{n_z}[n_x(x_t - c_x) + n_y(y_t - c_y)] \tag{5.32}$$

5.5.3.4 Curved Surface Contact

With relatively large offset and a small tilt angle, the curved surface contact occurs. It is assumed not to happen during the peg-in-tube assembly. Hence, its analysis is not included here. However, for the sake of completeness of the peg model, it is briefly being mentioned here. Any point $\mathbf{q}$ on the peg's curved surface may be assumed to be an equidistant point from the cylinder's axis. This may be represented as a point-line distance in 3D [30] as

$$r = \frac{|\mathbf{n} \times (\mathbf{c} - \mathbf{q})|}{|\mathbf{n}|} \tag{5.33}$$

With the scalar equation, Eq. (5.33), and conditions for the end caps boundaries, i.e., the plane equations, Eq. (5.21), one can detect any point falling inside the peg surfaces.

5.6 Depth-Profile-Based Search

The algorithm that makes use of the fact that when a tilted peg attains a two-point contact during rotation, the projection of the peg axis on the tube's top surface represents the direction of the hole and the peg center reaches the minimum depth c_z. Thus, the method of finding the hole direction involves rotating the tilted peg about the axis which is perpendicular to the tube's top plane and passes through the peg center. The peg rotates by one complete revolution and finally finds the angle α for which the peg lowers to the minimum depth c_z measured along the axis of rotation. This corresponds to the hole direction. The contact with the tube was maintained using the force control algorithm discussed in this chapter. Such a hole search process with a tube will have two minima in the depth profile. The proposed algorithm eliminates the minimum value which corresponds to the direction that will lead to peg being moved in the opposite direction to that of the actual hole, by checking the continuity condition of the depth profile at the minimum point. Algorithm 5.6.1 demonstrates how an analytical depth profile can be generated using the conditions discussed in Sect. 5.5.3. This was used later in Sect. 5.7. Figure 5.16 shows the analytical depth profile for a peg radius of $r = 9.42$ mm, tilt angle of $\theta = 7.5°$, and tube radius $R_1 = 9.7$ mm and $R_2 = 12.65$ mm. The two minima which can be observed in Fig. 5.16 lies opposite to each other, i.e., 180° apart at 134 and 314°. The inner tube hole will form a narrow depression in the profile such that the lowest point is non-differentiable. The profile with the wider opening and continuous at minima is formed when the flat bottom part of the peg rolls on the outer periphery of the tube. As the offset of the peg increases from the center of the hole, the depth realized for inner tilt which directs to the hole direction decreases and the depth for outer tilt which directs opposite to the direction of the hole increases. Hole direction can be safely detected just by checking the maximum depth location until the offset for which the depth due to inner tilt is greater than for the outer tilt. This is derived next.

Algorithm 5.6.1. GENERATE DEPTH PROFILE(θ, $\mathbf{c}$)

---

comment: Generate the depth profile

$r, R_1, R_2 \leftarrow$ **by definition**

for $\alpha \leftarrow 0$ **to** 360

do
- $\psi, \phi \leftarrow$ From $Eq.$ (5.24)
- $\mathbf{Q}_x \leftarrow function\_of(\psi)$
- $\mathbf{Q}_y \leftarrow function\_of(\phi)$
- $\mathbf{u} \leftarrow \mathbf{Q}_x\mathbf{Q}_y\mathbf{u}, \ \mathbf{v} \leftarrow \mathbf{Q}_x\mathbf{Q}_y\mathbf{v}$
- $\mathbf{n} \leftarrow \mathbf{u} \times \mathbf{v}$
- $P_l \leftarrow function\_of(\alpha)$ From $Eq.$ (5.19)
- **if** $x_l^2 + y_l^2 - R_1^2 < 0$
  - **then**
    - Solve $Eq.$ (5.22) and $Eq.$ (5.25) for P_i
    - $P_i \leftarrow P_i$ nearest to P_l
    - $c_z \leftarrow$ from $Eq.$ (5.27)
- **else if** $(x_l^2 + y_l^2 - R_1^2 > 0)$ and $(x_l^2 + y_l^2 - R_2^2 < 0)$
  - **then** $c_z \leftarrow r\sin(\theta)$ from $Eq.$ (5.28)
- **else if** $x_l^2 + y_l^2 - R_2^2 > 0$
  - **then**
    - $P_t \leftarrow$ From $Eq.$ (5.30)
    - **if** $E < 0$ for P_t. From $Eq.$ (5.25)
      - **then** $c_z \leftarrow$ from $Eq.$ (5.32)
      - **else**
        - Solve (5.22) with R_2 and $Eq.$ (5.25) for P_i
        - $P_i \leftarrow P_i$ nearest to P_l
        - $c_z \leftarrow$ from $Eq.$ (5.31)
- **save** $\alpha, \ c_z$

5.6.1 *Maximum Offset for Safe Hole Detection*

Figure 5.17a shows two-point contact case when the depth c_z is minimum for the inner tilt. Figure 5.17b, on the other hand, shows the outer rim contact where the depth c_z is again minimum. This is in the case of outer tilt. Let the offset be c_y measured along Y-axis, i.e., for $c_x = 0$, where the depth due to inner tilt is same as that of the outer tilt with the peg having tilt angle θ. The conditions for both the cases are as follows:

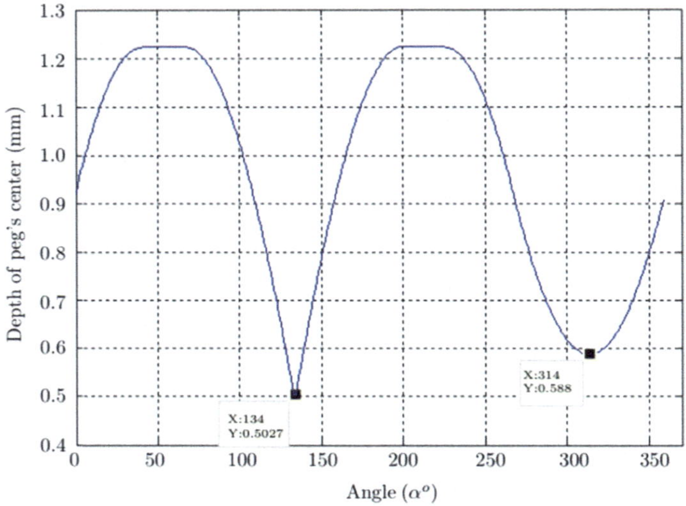

Fig. 5.16 An analytical depth profile

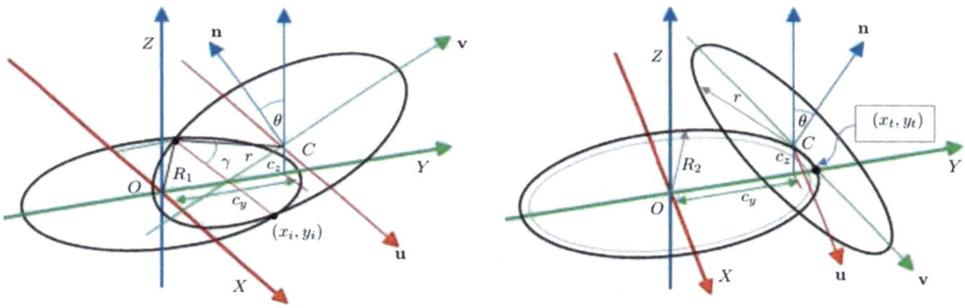

(a) Two point rim-rim contact during inner tilt

(b) Single point rim-face contact during outer tilt

Fig. 5.17 Instances of peg-in-tube contacts having same depth

$$
\begin{aligned}
&\text{(a) Inner tilt} \begin{cases} n_x = 0, \; n_y = -\sin\theta, \; n_z = \cos\theta \\ \quad \text{Contact point: } x_i, \; y_i \end{cases} \\
&\text{(b) Outer tilt} \begin{cases} n_x = 0, \; n_y = \sin\theta, \; n_z = \cos\theta \\ \text{Contact point: } x_t = 0, \; y_t = R_2, \; z_t = 0 \end{cases}
\end{aligned} \tag{5.34}
$$

Substituting the cases (a) and (b) in Eq. (5.34) in Eqs. (5.27) and (5.32), respectively, we get

$$
\begin{aligned}
\text{(a) } c_z &= \frac{1}{\cos\theta}\left[\sin\theta(R_2 - c_y)\right] \\
\text{(b) } c_z &= \frac{1}{\cos\theta}\left[-\sin\theta(y_i - c_y)\right]
\end{aligned} \tag{5.35}
$$

Using trigonometrical relations in Fig. 5.17(a), we get

$$r \sin\gamma = \frac{c_z}{\sin\theta} \text{ and } cos\gamma = \frac{x_i}{r} \tag{5.36}$$

The point of contact $(x_i,\ y_i)$ can now be obtained as

$$\begin{aligned} &\text{From Eq. (5.36), } x_i = r\sqrt{1 - \frac{c_z^2}{r^2\sin^2\theta}} \\ &\text{and from (5.35b), } y_i = c_y - \frac{c_z}{\sin\theta}\cos\theta \end{aligned} \tag{5.37}$$

Substituting $(x_i,\ y_i)$ from Eq. (5.37) in the condition $x_i^2 + y_i^2 = R_1^2$, we get

$$r^2 + c_y^2 - c_z^2 - \frac{2c_y c_z}{\tan\theta} = R_1^2 \tag{5.38}$$

Substituting c_z from (5.35a), a quadratic equation in c_y was obtained as

$$(3 - \tan^2\theta)c_y^2 - 2R_2(1 - \tan^2\theta)c_y + (r^2 - R_2^2\tan^2\theta - R_1^2) = 0 \tag{5.39}$$

Solving Eq. (5.39) for a realistic case with positive solution, c_y was obtained. This offset is expected to assist one to choose a suitable precision of the sensing system to approach the tube, i.e., a vision system, a laser range sensor, a 3D point cloud scanner, etc.

5.6.2 Hole Direction and Insertion

Once the rotation was completed for $0 \le \alpha \le 2\pi$, the value of α corresponding to the least depth $\alpha_{\min}$ was recorded. The peg was then made vertical, i.e., with tilt angle $\theta = 0$, and a constant desired contact force was maintained. The direction of the hole was obtained as

$$\Delta x = k_1 \cos\alpha_{\min} \text{ and } \Delta y = k_1 \sin\alpha_{\min} \tag{5.40}$$

or by using Eq. (5.24) as

$$\Delta x = k_2 \sin\psi_{\min} \text{ and } \Delta y = k_2 \sin\phi_{\min} \tag{5.41}$$

where k_1 and k_2 are the constants of proportionality. Figure 5.18 shows the top view of the two-point contact case where the depth is minimum during the search procedure. The axis of the peg, when projected on the top surface of the tube, passes through the hole center.

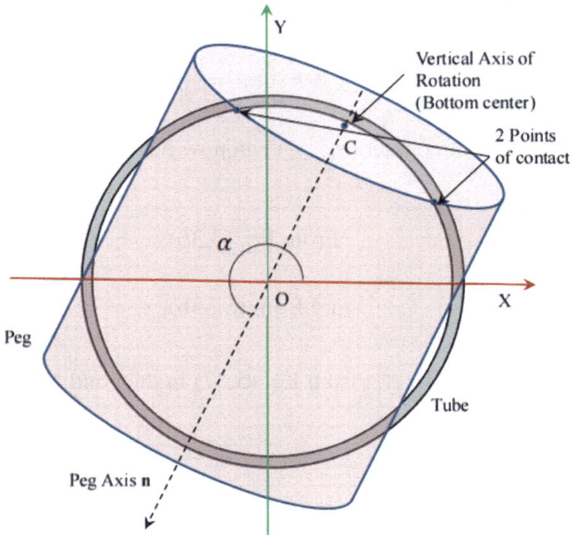

Fig. 5.18 Top view of the two-point contact case of peg-in-tube

Now, the peg was advanced with continuous small displacements of Δx and Δy along X and Y, respectively, until it senses a reduced reaction force due to the presence of the hole. The robot was now stopped and the peg was inserted gradually into the hole. Note that the values of k_1 or k_2 directly affect the advancing velocity toward the hole. Their value depend on the clearance between the tube and the hole. They were kept small, i.e., 0.004 in our experiments due to the possibility of skipping of the peg over the hole.

5.6.3 *Implementation*

In order to use the proposed algorithm, the steps given in Algorithm 5.6.2 were used with a KUKA KR5 Arc robot. It was assumed that the peg is to be brought near the top of the tube using any suitable system, for example, a vision system. The robot provides the end-effector coordinates (X, Y, Z) in its base frame, where Z is vertically up and parallel to the tube axis. The peg was tilted and lowered using a force control mode, i.e., if the peg establishes any contact it would maintain the desired force $F_{desired}$. The small decrement in the robot's end-effector coordinate Z, i.e., Δz, should be kept very small as establishing a contact at high speed will give rise to an undesired thrust that can damage the surface of tube or peg. While the peg is rotated for $0 \leq \alpha \leq 360°$ the algorithm updates the minimum Z value, i.e., $Z_{\min}$ with the current Z if it is lower than the existing $Z_{\min}$. It also sets the variable $\alpha_{\min}$ to the value of α corresponding to $Z_{\min}$. Once the rotation is complete the peg was made vertical by setting the tilt angle θ to 0. The peg was then moved toward the hole using Eq. (5.41) till it finds the hole, where the peg senses a reduced reaction force F_z than the desired force, or the peg gets into the hole by a small distance d. The peg was then lowered gradually into the hole.

Algorithm 5.6.2. PEGINTUBE(θ, **c**)

---

comment: Algorithm for implementation

START: Peg over the tube
$Z_{min} \leftarrow Z$
$\theta \leftarrow$ **tilt angle**
while $F_z \leq F_{desired}$
  do $Z \leftarrow Z - \Delta z$
for $\alpha \leftarrow 0$ **to** 360°
$$\textbf{do} \begin{cases} \psi, \ \phi \leftarrow \text{From (5.24)} \\ \textbf{if } Z < Z_{min} \\ \quad \textbf{then} \begin{cases} Z_{min} \leftarrow Z \\ \alpha_{min} \leftarrow \alpha \end{cases} \end{cases}$$
$\theta \leftarrow 0$
$\Delta x, \ \Delta y \leftarrow$ From (5.41)
while $F_z > F_{desired}$ **OR** $Z > Zmin - d$
$$\textbf{do} \begin{cases} X \leftarrow X + \Delta x \\ Y \leftarrow Y + \Delta y \end{cases}$$
END: Insert the peg

5.7 Experimental Results of Peg-in-Tube Task

The KUKA KR5 Arc industrial robot was used to validate the proposed peg-in-tube algorithm. The parameters for the tube and for the peg were same as in Sect. 5.6. A monocular camera system was used to bring the peg directly over the tube. The end effector was mounted with a six-component force/torque sensor manufactured by SCHUNK of type Delta, SI-165-15. In order to measure the vertical depth during the hole search procedure as discussed in Sect. 5.6, we relied on the in-built forward kinematics of KUKA KR5 Arc robot. It was assumed that the grasping was done precisely every time a new peg was picked. Figure 5.19 shows the depth profiles obtained experimentally and analytically for peg radius of $r = 9.42$ mm, tilt angle of $\theta = 7.5°$, tube radii $R_1 = 9.7$ mm and $R_2 = 12.65$ mm, and an offset of $\mathbf{c} = (-7.53, \ -7.2, \ 0)$ mm. A close match of the depth profiles validates the proposed Algorithm 5.6.1. The difference in the depth profiles is mainly due to experimental procedure. These are discussed in the following sub-sections.

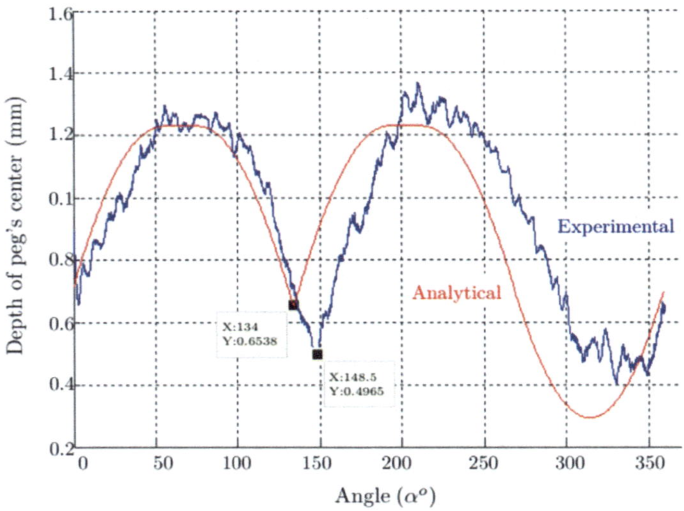

Fig. 5.19 Theoretical versus analytical depths

5.7.1 *Force/Torque Sensor and DAQ*

The force/torque sensor used in the experiment was of high bandwidth with $495N$ along vertical Z-direction and a low sensor resolution of $1/16N$ using a 16-bit Data Acquisition (DAQ) system. The DAQ was used to pass the analog voltages of 6 strain gauges to the controller. These data were filtered using a real-time low-pass fourth-order Butterworth filter with cutoff frequency of 40 Hz, based on the sampling frequency of 83.33 Hz. The force data was generated by multiplying the six-dimensional vector of voltages with the 6×6 sensor calibration matrix. The resulting force data had a typical noise of $\pm 0.15N$ which is evident from Fig. 5.20 that shows the actual acquired data for two different instances along with the filtered data in dark.

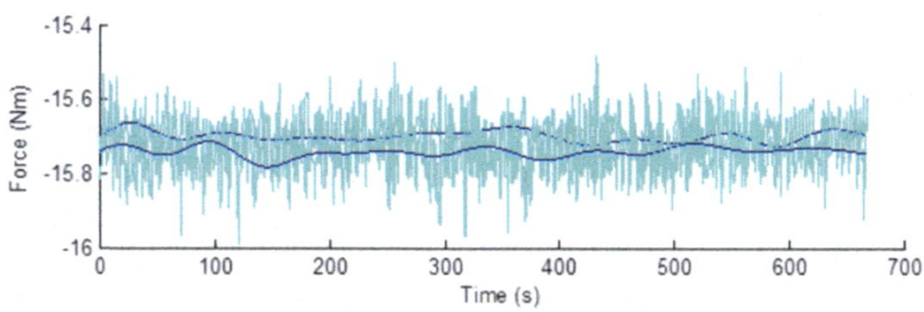

Fig. 5.20 Data acquired for two different instances

5.7.2 *Effect of Peg Size and Clearance*

The bigger peg will definitely accommodate more placement error, i.e., the offset, in order to have a safe hole direction detection as per Eq. (5.39). However, in order to study the effect of peg sizes, two different pegs were analyzed. Figure 5.21 shows the analytical depth profile for two different pegs of radii 9.42 and 24.0 mm when the peg center lies on the tube rim, i.e., with offset equal to the mean tube radius $c_x = (R_1 + R_2)/2$ and $c_y = 0$. The peg and tube had a radial clearance of 0.28 mm for both the cases.

It was observed that the differences in depths of two minimums are 0.526 and 1.516 mm, respectively, for smaller and bigger pegs. These differences are required to be distinguished clearly using depth observations made by robot's sensory system to estimate the correct hole direction. It becomes even more difficult when the depth data are accompanied with the noise as well. This limits the minimum size of the peg which can be handled using the proposed algorithm with the setup used in this work. Peg of diameter 18 mm with 95% was successfully inserted with success rate (tested with 40 pegs) using the existing setup. For smaller pegs and smaller offset, a blind spiral search [18] may be used instead.

It may be observed in Fig. 5.21 that an error in depth measurement by $0.1mm$ will cause corresponding hole direction error of 6 and 2° for smaller and the bigger pegs, respectively. This quantifies the hole direction error for a particular depth measurement error. If the smaller peg proceeds with directional error of 6° with offset amounting to the peg radius, it will reach to 0.984 mm away from the actual

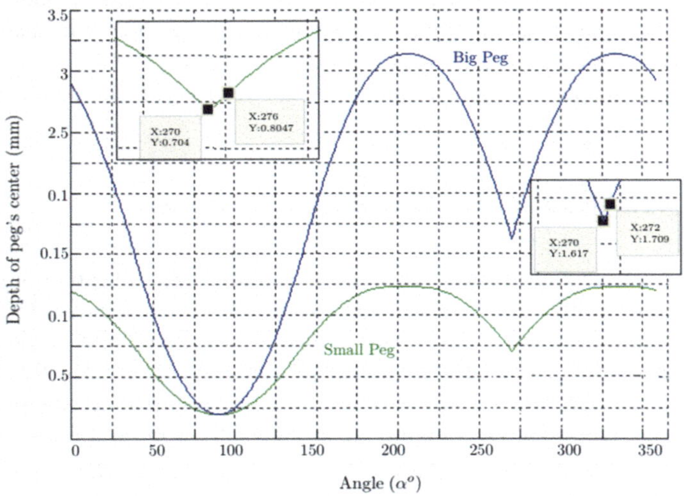

Fig. 5.21 Depth profile obtained by varying angle for two sizes of peg, namely, 24.5 mm and 28 mm

hole. In the case of bigger peg, 2° directional error will lead to 0.977 mm hole position error. Thus, a system with an error of 0.1 mm in depth measurement may not successfully perform peg insertion with a peg and tube having clearance of 1 mm or less.

5.7.3 Other Sources of Error

Apart from force sensing and displacement measurement errors, this section lists additional sources of errors that limit the performance of the peg-in-tube process. Following points may be noted while designing the peg-in-tube assembly systems:

- Passive compliance of the gripper system allows unaccounted motion of the peg-tip.
- Assuming the tube's top face perfectly flat is just an ideal case. Peg and tube axes alignment is also required which may be done using [21].
- Poor surface finish of the rolling surfaces of peg and tube creates force/displacement noise.
- Two-finger gripper cannot hold the peg firmly enough to restrict small linear and angular motions of the peg with respect to the robot's end effector during contact.
- Due to inaccurate tool calibration, the peg never makes pure rolling during the search procedure. Any slippage on the edge will create noise.
- At high-speed force control response is poor which may cause undue contact forces or a loss of contact. However, with 10% point-to-point speed of KUKA KR5 Arc robot four consecutive pegs were inserted in 22.5 seconds. This included picking of peg from a fixed location, bringing the peg directly over the tube, searching for the hole, and finally inserting the peg, for each of the peg.
- An industrial floor with high vibration may also lead to reduced performance.

5.7.4 Force Versus Depth-Based Hole Detection

In order to have a comparison between the depth-based localization proposed here and the force-based localization, both the depth and the forces/moments were recorded during the rotation of the peg. The radius of the tube was 23.62 mm, while the inner and outer radii of the tube were 24.5 mm and 28 mm, respectively. The tilt angle was set to 7.5°. The offset was kept so as to have just one minima, i.e., the peg's outer rim falls on the tube's flat face. Figure 5.22a shows the polar plot for the variation of the depth with respect to angle α. The variation of vertical force can be seen in Fig. 5.22b. The coherence of force plot with the depth plot where the minimum depth may be observed is notable. A sharp rise in force was observed that can be effectively used for hole direction estimation. This surge force was created when the peg starts rising up after the two-point contact, i.e., the minimum depth. The hole direction

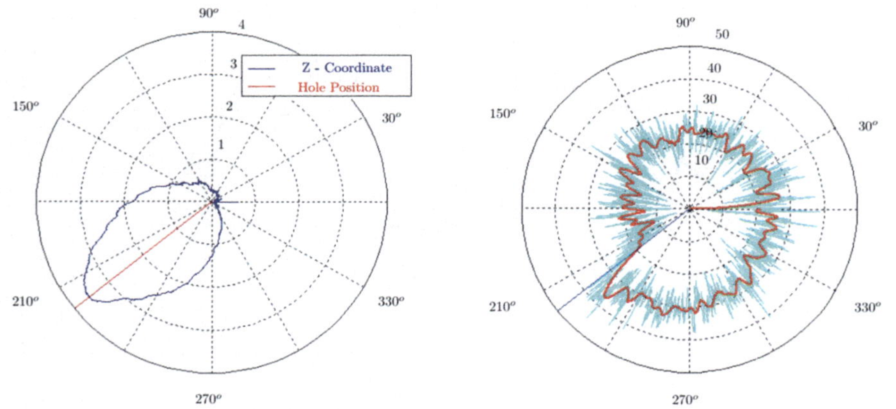

(a) Depth profile (b) Force profile

Fig. 5.22 Profiles for the peg of radius 23.62 mm

due to force profile is consistent to that of the depth profile, i.e., at $\alpha = 218.4°$. The force data shows the filtered data in dark, along with the actual data acquired through the force sensor. With the unfiltered data it is difficult to figure out the surge due to the hole in real time, whereas with filtered data a lag was observed that might again lead to a wrong hole direction. One may estimate the lag with proper knowledge of the filter parameters. It may, however, be inferred with the data that is shown in Fig. 5.22 that the depth data provides a better localization of the hole direction. A similar surge in moments was also observed at $214.8°o$, i.e., at the hole direction as shown in Fig. 5.23.

There are various checks which one may devise in order to encounter any failed attempt and incorrect placement of the peg. For example, one may check if the peg moves more than its radius while moving toward the hole. Accordingly, the robot algorithm may be designed to search the hole again. Similarly, if the peg gets into the hole without touching the tube surface, the peg may be released without starting the search procedure. The system may also be programmed to remember the hole location once it is located using the proposed search technique to repeatedly insert the peg at the same hole location. Again as a check, one may lower the peg using a *guarded move*, i.e., by continuously checking the contact force, and if the peg hits the tube surface the search process may be redone.

5.7.5 Results of the Framework

In order to validate the data generated by the proposed framework, a cylindrical peg of diameter 19.0 mm was used with the hole diameter of 19.6 mm. The CAD models

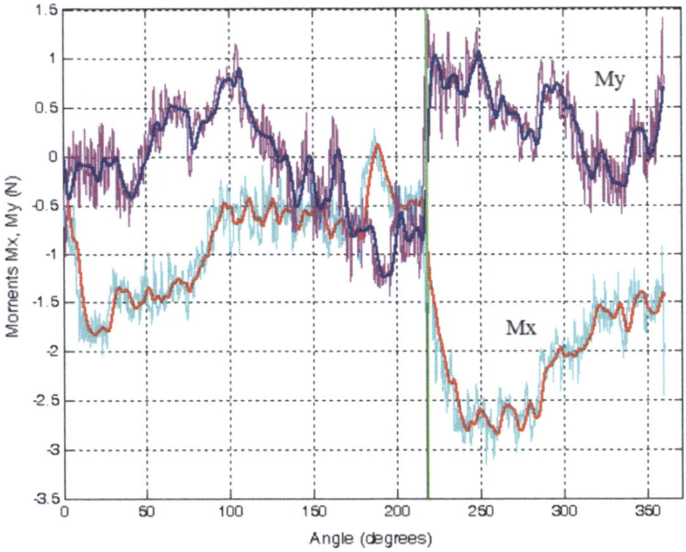

Fig. 5.23 Moment profile for the peg of radii 23.62 mm

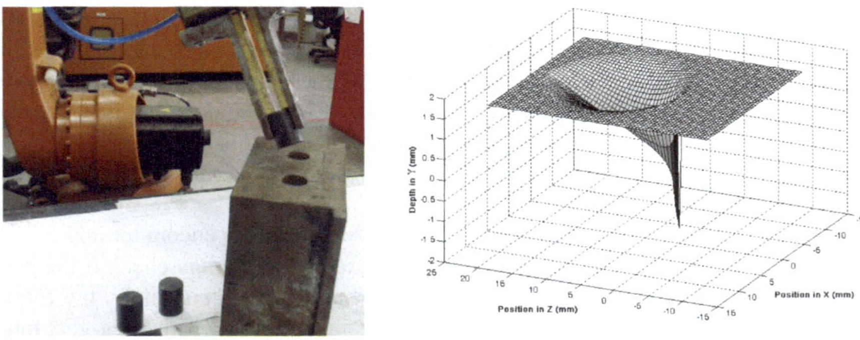

(a) Actual Peg-in-Hole setup

(b) A sample depth-map for cylindrical peg and hole

Fig. 5.24 Actual Peg-in-Hole setup and depth profile obtained experimentally

of the peg and the hole block were prepared with the same size as that of an existing experimental setup shown in Fig. 5.24a. In Fig. 5.24a, a robot is holding the peg using its gripper and the hole block is lying below. Figure 5.24b shows a sample depth-map generated when the peg was tilted with an angle of 10°. The steps for increments along XZ-plane, i.e., the hole surface taken here, were taken as 0.5 mm and the resolution for depth along Y was taken as 0.1 mm.

This result is similar to the one which was shown in [17] for the cylindrical peg. However, their peg diameter was different, which was 100 mm. In order to validate the accuracy of the depth-map generated with the proposed framework, a depth-map using analytical technique was also generated for the same set of the peg and hole, using the cases discussed in Sect. 5.5.3.

5.8 Summary

This chapter presents a current control-based approach for converting a existing joint into a passive compliant joint for an industrial robot. An active force control scheme is proposed with proportional controller and steady-state error compensator. Passive joint compliance by limiting the current with active force control is reported for the first time. The parallel active–passive force control scheme was tested for an effective force/position control along with compliant-like behavior of the links while robot–robot collision and robot–human interaction occur. Such a controller is expected to be intrinsically safe from potential pinching or trapping hazards and will also minimize the collision hazards. Such a controller may also be useful in cooperative robot manipulation task.

In order to prove the efficacy of the parallel active–passive force controller, this chapter presents a thorough geometrical analysis of the "peg-in-tube" assembly process and proposes a novel algorithm based on depth measurements of peg center to perform "peg-in-tube" task. The results are demonstrated on a KUKA KR5 Arc industrial robot with a chamferless cylindrical peg and a tube having a tight tolerance. The demonstration of the proposed parallel active–passive force controller may be seen at [31]. The application of transforming the joint into a compliance joint by limiting the current for the safer operation in a reconfigurable mobile robot is adopted in [32].

Please scan above for videos and refer Appendix C.3 for files detail

References

1. Zeng, G., Hemani, A.: An overview of robot force control. Robotica **15**(5), 473–482 (1997)
2. Schumacher, M., Wojtusch, J., Beckerle, P., von Stryk, O.: An introductory review of active compliant control. Robot. Autonomous Syst. **119**, 185–200 (2019)
3. Caccavale, F., Natale, C., Siciliano, B., Villani, L.: Integration for the next generation: embedding force control into industrial robots. IEEE Robot. Automat. Mag. **12**(3), 53–64 (2005)
4. Schutter, J.D., Brussel, H.V.: Compliant robot motion II. A control approach based on external control loops. Int. J. of Robot. Res. **7**(4), 18–33 (1988)
5. Khatib, O., Thaulad, P., Yoshikawa, T., Park, J.: Torque-position transformer for task control of position controlled robots. In: IEEE International Conference on Robotics and Automation, (ICRA 2008), pp. 1729–1734 (2008)
6. Schutter, J.D.: A study of active compliant motion control methods for rigid manipulators based on a generic scheme. In: IEEE International Conference on Robotics and Automation (ICRA 1987), vol. 4, pp. 1060–1065 (1987)
7. Okada, M., Nakamura, Y., Hoshino, S.: Design of active/passive hybrid compliance in the frequency domain-shaping dynamic compliance of humanoid shoulder mechanism. In: IEEE International Conference on Robotics and Automation (ICRA 2000), vol. 3, pp. 2250–2257 (2000)
8. Ham, R., Sugar, T., Vanderborght, B., Hollander, K., Lefeber, D.: Compliant actuator designs. IEEE Robot. Automat. Mag. **16**(3), 81–94 (2009)
9. Geravand, M., Flacco, F., Luca, A.D.: Human-robot physical interaction and collaboration using an industrial robot with a closed control architecture. In: IEEE International Conference on Robotics and Automation (ICRA 2013), pp. 4000–4007, Karlsruhe, Germany (2013)
10. Vick, A., Surdilovic, D., Kruger, J.: Safe physical human-robot interaction with industrial dual-arm robots. In: 2013 9th Workshop on Robot Motion and Control (RoMoCo), pp. 264–269 (2013)
11. Song, J., Chen, Q., Li, Z.: A peg-in-hole robot assembly system based on gauss mixture model. Robot. Comput. Integr. Manuf. **67**, 101996 (2021)
12. Vasic, M., Billard, A.: Safety issues in human-robot interactions. In: IEEE-RAS International Conference on Robotics and Automation (ICRA 2013), Karlsruhe, Germany (2013)
13. Chen, H., Li, J., Wan, W., Huang, Z., Harada, K.: Integrating combined task and motion planning with compliant control. Int. J. Intell. Robot. Appl. **4**(2), 149–163 (2020)
14. Kim, Y., Kim, B., Song, J.: Hole detection algorithm for square peg-in-hole using force-based shape recognition. In: 8th IEEE International Conference on Automation Science and Engineering, pp. 1074–1079, Seoul, Korea (2012)
15. Katsura, S., Matsumoto, Y., Ohnishi, K.: Modeling of force sensing and validation of disturbance observer for force control. IEEE Trans. Indus. Electron. **54**(1), 530–538 (2007)
16. Chhatpar, S.R., Branicky, M.S.: Particle filtering for localization and in robotic assemblies with position uncertanity. In: IEEE/RSJ International Conference on Intelligent Robots and Systems (IROS 2005), pp. 3610–3617 (2005)
17. Sharma, K., Shirwalkar, V., Pal, P.K.: Intelligent and environment-independent peg-in-hole search strategies. In: IEEE International Conference on Control, Automation, Robotics and Embedded Systems, pp. 1–6, Noida, India (2013)
18. Chen, H., Zhang, G., Zhang, H., Fuhlbrigge, T.A.: Integrated robotic system for high precision assembly in a semi-structured environment. Assembly Automation **27**(3), 247–252 (2007)
19. Raibert, M.H., Craig, J.J.: Hybrid position / force control of manipulators. ASME J. Dyn. Syst. Meas. Control **102**, 126–133 (1981)
20. Nemec, B., Abu-Dakka, F., Ridge, B., Ude, A., Jorgensen, J., Savarimuthu, T., Jouffroy, J., Petersen, H., Kruger, N.: Transfer of assembly operations to new workpiece poses by adaptation to the desired force profile. In: 16th International Conference on Advanced Robotics (ICAR 2013), pp. 1–7 (2013)

21. Huang, S., Murakami, K., Yamakawa, Y., Senoo, T., Ishikawa, M.: Fast peg-and-hole alignment using visual compliance. In: IEEE/RSJ International Conference on Intelligent Robots and Systems, pp. 286–292. Tokyo, Japan (2008)
22. Ma, D., Hollerbach, J.M.: Identifying mass parameters for gravity compensation and automatic torque sensor calibration. In: IEEE International Conference on Robotics and Automation (ICRA 1996), pp. 661–666 (1996)
23. KUKA: KUKA.RobotSensorInterface 2.3. KUKA Roboter GmBH, Germany (2009)
24. Hayat, A.A., Udai, A.D., Saha, S.K.: Kinematic-chain identification and gravity compensation of KUKA KR5 robot. In: 1st International and 16th National Conference on Machines and Mechanisms (iNaCoMM '13), IIT Roorkee (2009)
25. Saha, S.K.: Introduction to Robotics, 2nd edn. Mc-Graw Hill, New Delhi (2014)
26. Atkeson, C., An, C., Hollerbach, J.: Rigid body load identification for manipulators. In: 24th IEEE Conference on Decision and Control, vol. 24, pp. 996–1002 (1985)
27. Winkler, A., Suchy, J.: Position feedback in force control of industrial manipulators—an experimental comparison with basic algorithms. In: IEEE International Symposium on Robotic and Sensors Environments (ROSE 2012), pp. 31–36 (2012)
28. Udai, A.D., Hayat, A.A., Saha, S.K.: Parallel active/passive force control of industrial robots with joint compliance. In IEEE/RSJ Internatinal Conference on Intelligent Robots and Systems, Chicago, Illinois (2014)
29. Messner, W.C., Tilbury, D.: Control Tutorials for MATLAB®and Simulink®: A Web Based Approach. Pearson Education, Inc (1998)
30. Weisstein, E.W.: Point-Line Distance–3-Dimensional (2016). http://mathworld.wolfram.com/Point-LineDistance3-Dimensional.html. [Online; accessed 01-September-2014]
31. Udai, 2014: Video demonstrating the proposed parallel active/passive force controller (2014). https://www.youtube.com/user/arunudai. [Online; uploaded 21-June-2014]
32. Hayat, A.A., Karthikeyan, P., Vega-Heredia, M., Elara, M.R.: Modeling and assessing of self-reconfigurable cleaning robot htetro based on energy consumption. Energies **12**(21), 4112 (2019)

Chapter 6
Integrated Assembly and Performance Evaluation

This chapter focuses on applying the algorithm developed for the multiple sensors calibration mounted on the robot, as discussed in Chap. 2, with the accounting for the sensitivity analysis of the parameters in Chap. 4. Moreover, having only the vision system is not sufficient for the peg-in-tube task's precision-based assembly task. The model-based control approach was implemented incorporating the identified kinematic and dynamic parameters presented in Chaps. 4 and 5, respectively, to address this issue. The detailed approach for the vision based is explained. The pick-up was the cylindrical pellets randomly kept inside a bin with occlusion which is explained. The pick-up was done using the information generated by the sensor fusion, and placement of it is near the tube with a clearance of the order 10^{-5} m. The search algorithm's performance integrating the passive–active control technique explained in Chap. 5 is extensively tested in experiments for the insertion of pellets, i.e., peg in the tube.

Figure 6.1 shows the topics related to the assembly task mainly for the peg-in-hole/tube summarized using over 200 literature published. The integration of multiple technologies in Fig. 6.1 highlights the knowledge base required for the demonstration of the assembly task. The integrating vision-based pick-up with force–torque sensor-based assembly helps us to build a complete system. Robotic pick-up and assembly is a challenge for any industry. In environments like nuclear power stations and similar environments where it is required to pick up and insert the nuclear fuel rods into a pipe, and other dangerous and repetitive task such an automated system will eliminate human beings from doing unsafe tasks.

Supplementary Information The online version contains supplementary material available at https://doi.org/10.1007/978-981-16-6990-3_6.

A. A. Hayat et al., *Vision Based Identification and Force Control of Industrial Robots*,
Studies in Systems, Decision and Control 404,
https://doi.org/10.1007/978-981-16-6990-3_6

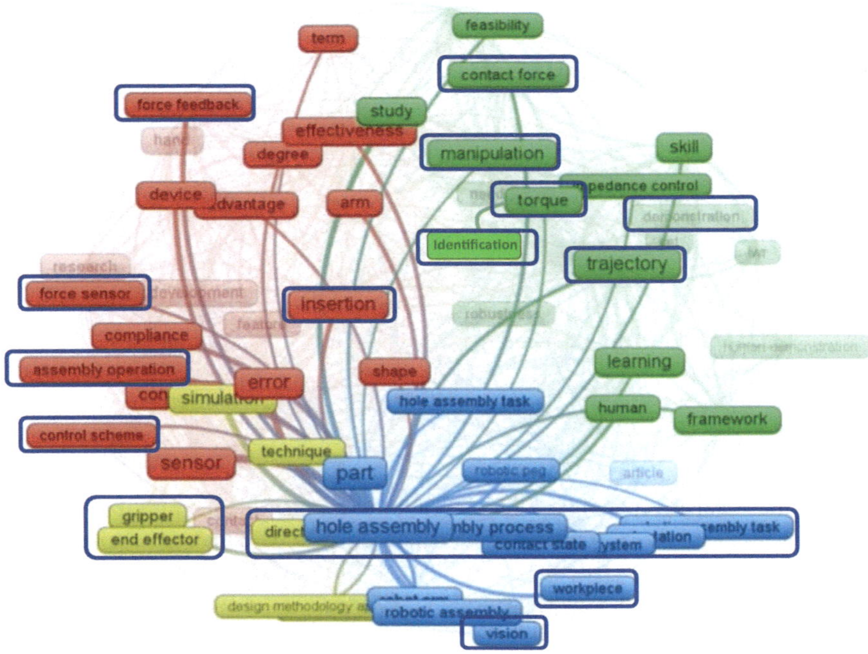

Fig. 6.1 Essential components and its connectivity for the assembly task

6.1 Integration Setup

The first step of the algorithm involves identifying the location of the pellet through the vision sensor. The camera is moved to the home position to capture the image in its field of view, from which it first identifies one of the pellets in the image and then detects its position. The methods already discussed in Chap. 2 are used. The integration of the sensor-based pick-up of pellets and control-based insertion of pellets inside the tube is depicted in Fig. 6.2. The robot is equipped with the mounting of sensors and gripping tools, namely, a monocular camera, force/torque (F/T) sensor, two-fingered gripper, and a suction gripper. A camera and a laser range sensor act as input for our setup. The camera is from Basler of resolution 2448 × 2048 for image scan and micro-epsilon high-accuracy scanner for the 3D scanner. Both sensors are mounted on the robot's end effector. The algorithm runs on a 2.8 GHz Intel Xeon CPU.

The flowchart for the bin-picking and insertion of pellets inside the tube in the two orientations, namely, horizontal and vertical, is shown in Fig. 6.3. The estimation of the pellets location and the tube hole is done using a camera mounted on the end effector of the KUKA KR5 Arc. Moreover, the active/passive control loop is active throughout the process of pick-up and insertion task. The method for the

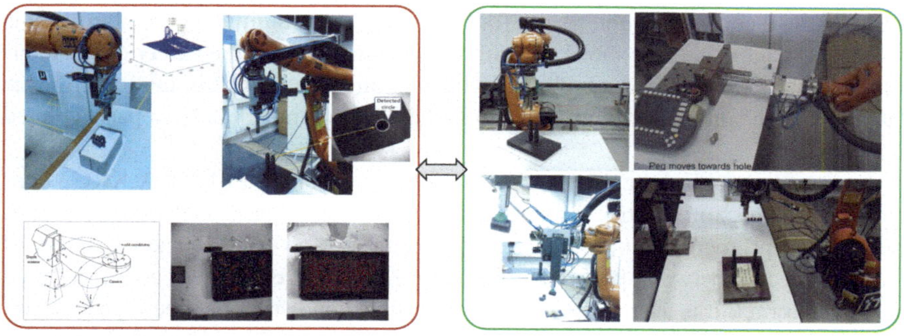

(a) Vision based object detection and calibration

(b) Control of KUKA KR5 Arc with its identified kinematic and dynamic models

Fig. 6.2 Integration of Vision and Force/Torque

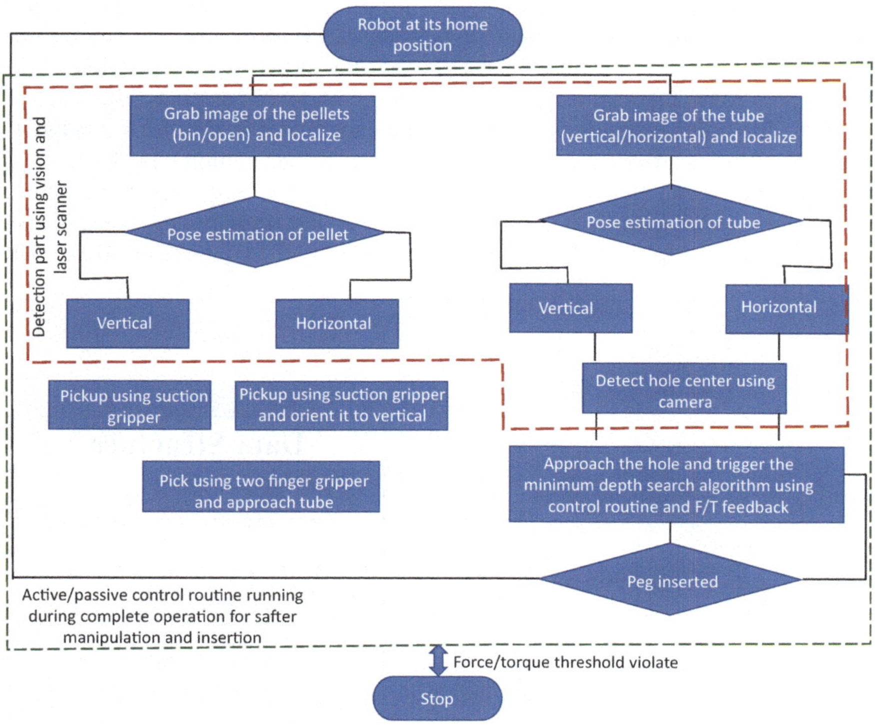

Fig. 6.3 Integration of the sensory-based pick-up and active/passive control-based insertion of pellets

active/passive control and the search for hole using the minimum depth algorithm is detailed in Chap. 5. The control algorithm presented ensures the safety of the environment and the robot in case of any collision, which will result in the force/torque data spike or the motor current will result in the robot to immediately stop as depicted in Fig. 6.3.

The communication between different computers and with KUKA KRC2 controller: Transmission Control Protocol (TCP) of Internet Protocol (IP) suite is one of the standard protocols that provides reliable, ordered, and error-checked delivery of a stream of data between programs running on different computers connected in a LAN or Internet. It is used only after establishing a handshake between the computers where the exchange of data occurs. On the other hand, User Datagram Protocol (UDP), another protocol of IP suite, can be used to send messages, referred to as datagrams, between programs and applications without establishing a prior communication between the computers. UDP is used primarily to broadcast data over the network as compared to TCP, which is used for dedicated data transmission. In this project, TCP was used to communicate between the KUKA KRC2 controller and the computer. Through this, the robot was controlled using RSI (the software from KUKA permitting communication between robot and computer), as illustrated in Fig. 6.4. The force/torque sensor used for the experiment was of high bandwidth with $495N$ along vertical Z-direction and a low sensor resolution of $1/16\ N$ using a 16-bit Data Acquisition (DAQ) system. The DAQ was used to pass the analog voltages of 6 strain gauges to the controller. These data were filtered using a real-time low-pass fourth-order Butterworth filter with cutoff frequency of 40 Hz. The

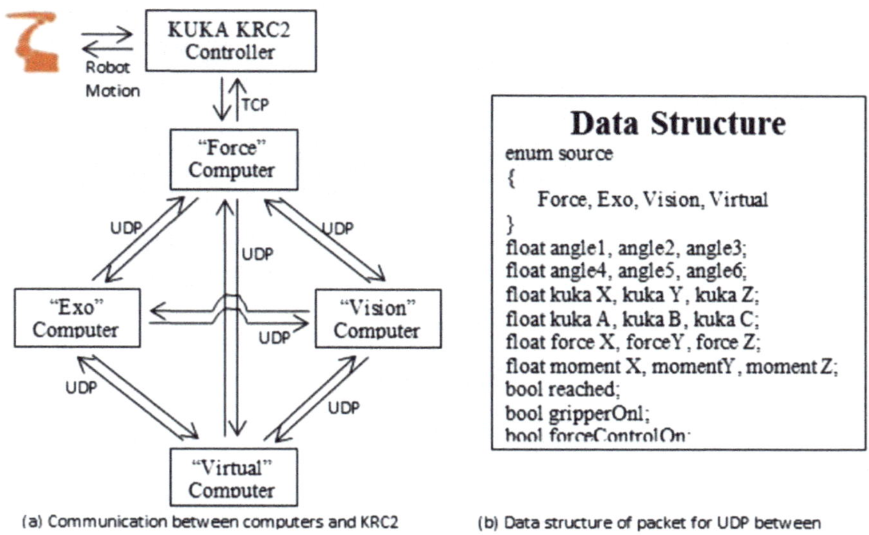

Fig. 6.4 TCP and UDP communication between computers and controller

sampling frequency was of 83.33 Hz. The force data was generated by multiplying this six-dimensional vector of voltages to the 6×6 sensor calibration matrix. The resulting force data had a typical noise of $\pm 0.15N$.

6.2 Bin Picking of Pellets

In an industrial setting, there might be a requirement to assemble objects in other fashions rather than what is discussed earlier. It is commonly required that after picking up the desired object, a robot may have to load a machine with these objects with no human intervention. This section discusses the method to load machines with such objects using a computer program. Our assumption for this experiment is that the input receptor of the machine is metallic tubes.

Cluttered pellets in multi-layer configuration introduce several challenges, unlike the situation in which single-layer configuration allowed assumption of the pellet height, which was used to calculate the coordinates in 3D space. Using an accurate range sensor was thus opted and is shown in Fig. 6.5a.

In multi-layer configuration and having occlusion, this height assumption does not hold. Thus, a different type of sensing is required over the conventional single image used. A depth sensor in multi-layer configuration becomes a necessity. However, it can be argued that a multi-view camera or stereo image could be used for creating such a depth-map. Image-based reconstruction heavily relies on the presence of contrast and features born out of such contrast. Our setup is decisively lacking in contrast, i.e., the black pellets in the black configuration would challenge the reliability of such techniques, and thus image-based reconstruction is not a viable option. The sensor provides us with a high-resolution scan creating a 3D point cloud. Besides resolving the issue of layers, it inherently models occlusion very well due to depth data. With this sensor, it was possible to develop an algorithm that accurately determines the pellet position in 3D space with varying height. Our algorithm starts with the segmentation of 3D data gathered from the boat to isolate connected patches of similar curvature and orientation. Since the resolution of scan is limited multiple pellets aligned in a similar configuration may come out to be in the same patch. To distinguish between these, we use information obtained from image in this region in correspondence to our results from 3D segmentation. The image data is also used to keep track of current bin configuration to monitor any disturbance in between scans. The first step involves identifying the location of the pellet through the vision and laser scanner. The proposed method for the detection of pellet and pick-up from the bin in a given settings is discussed next.

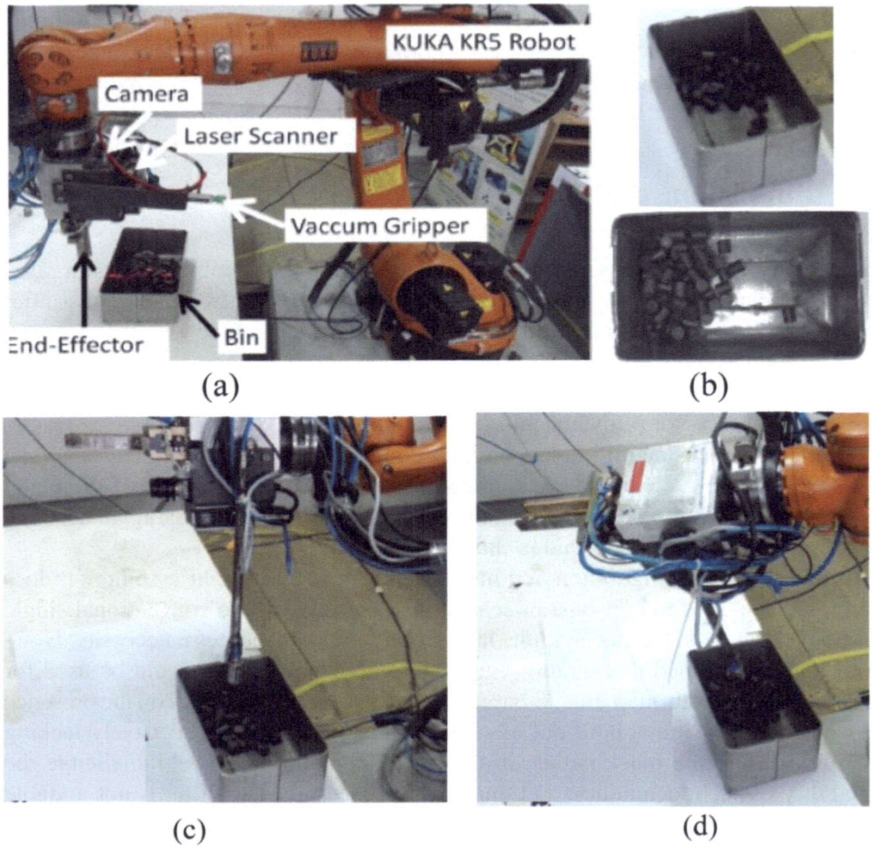

Fig. 6.5 Bin picking of pellets using the suction gripper

6.2.1 Calibration

The calibrated camera along with the transformation between the 2D laser scanner and camera is used. The steps and the proposed method using a monocular camera are detailed in Chap. 2.

6.2.2 Segmentation

The segmentation process begins with localization of the point cloud and finding a patch corresponding to a single cylinder. If the size of the patch is bigger than the size of gripper diameter, it is qualified for further processing else it is dropped as it considered to be unapproachable. Effort was then made to find geometric features of the patch, concluding in type of surface (cylindrical or flat), centroid, approach normal, and axial orientation. Also estimated are the length and breadth of the patch in the object frame. If the length and breadth of the patch are within acceptable deviation from the standard size of the pellet, we qualify the computation, commanding the robot to pick up the pellet. Else we perform localized image segmentation based on a context generated by information from 3D segmentation. If the patch size is smaller than expected we search for a single pellet, else multiple pellets are extracted. In this fashion, information from the two sensors is used to complement each other. We are utilizing the strength of each form of segmentation to provide improved results.

6.2.3 Collision Avoidance

To pick up objects scrambled in a box is a straightforward task for humans, though its automation requires multiple considerations. To achieve this automation, it becomes necessary that the system thus formed is free from collision and can pick up the object specified. To deal with this problem, a solution has been modeled based on the geometry of the bin and the mechanical arrangement of our setup. The collision avoidance of the extended rod for the vacuum gripper as per the experimental setup is shown in Fig. 6.5b, c. It has to be accounted for during the pick-up task from the bin. The formulation for it is presented next.

Arbitrary and complex orientations increase the likelihood of collision between the gripper and container walls. Therefore, before attempting to pick up, we ensure the gripper's required orientation would not result in a collision with the container. In this section, we define an analytical approach which is inspired from [1] where cylinder-to-cylinder interaction was modeled. Our approach is different as we model cylinder and plane collision by approximating the gripper rod with a cylinder and a container's walls by finite planes. The cylinder is assumed to be oriented with axis aligned along the surface normal of the object, given by $\left(n_x, n_y, n_z\right) \in \mathbb{R}^3$ and with the center of face of the cylinder coincident with patch surface centroid, given by $\left(p_x, p_y, p_z\right) \in \mathbb{R}^3$. It is then analyzed for proximity to planes. The analysis begins by projecting the system's geometric model onto the xy-plane, which is parallel to the table. The steps are as follows.

Find the planes in direction of orientation of the cylinder that can possibly suffer collision. Let $f : \mathbb{R}^2 \times \mathbb{R}^2 \rightarrow \{(AB, BC)\,(BC, CD)\,, (CD, DA)\,, (DA, AB)\}$ where AB, BC, CD, and DA represent planes projected as lines in the XY-plane as shown in Figs. 6.6a and 2.7b. Note that edge ABCD is equivalent to EFGH, Fig. 2.7b.

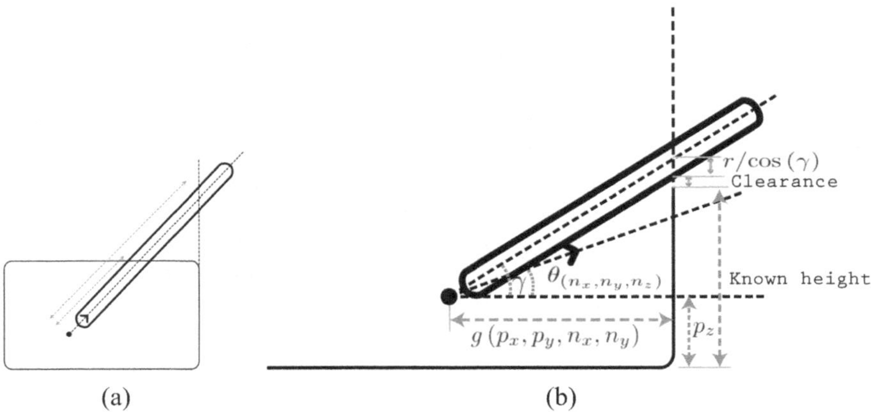

Fig. 6.6 **a** XY-plane (top view), **b** cylinder-axis and Z-axis plane (side view)

$$f\left(n_x n_y\right) = \begin{cases} (AB, BC) & 90 > \tan\left(\frac{n_y}{n_x}\right)_i \geqslant 0 \\ (BC, CD) & 180 > \tan\left(\frac{n_y}{n_x}\right)_i \geqslant 90 \\ (CD, DA) & 270 > \tan\left(\frac{n_y}{n_x}\right)_i \geqslant 180 \\ (DA, AB) & 360 > \tan\left(\frac{n_y}{n_x}\right)_i \geqslant 270 \end{cases}$$

To calculate distance of both walls from the pick point along pick up orientation, let $f\left(n_x n_y\right) = (AB, BC)$ and $f\left(n_x n_y\right)(1) = AB$, similarly $f\left(n_x n_y\right)(2) = BC$. Now let $d\colon \mathbb{R}^2 \times \mathbb{R}^2 \times \mathbb{R} \to \mathbb{R}$, where $d\left(\left(p_x, p_y\right), \left(n_x n_y\right), i\right) =$ distance of point (x,y) from along direction $\left(n_x, n_y\right)$ from line $f\left(\left(p_x, p_y\right), \left(n_x, n_y\right), (i)\right)$. Then the distance pick-up point from the wall closest along the cylinder axis is given by $g\colon \mathbb{R}^2 \times \mathbb{R}^2 \to \mathbb{R}$, where

$$g\left(\left(p_x, p_y\right), \left(n_x, n_y\right)\right) = min_{i=1,2}\left(d\left(\left(p_x, p_y\right), \left(n_x n_y\right), i\right)\right).$$

For the following steps, we now restrict the discussion to the plane defined by the cylinder axis and the projected normal given by the line $\left(p_x, p_y, p_z\right) + t\left(n_x, n_y, 0\right)$. Let the overall angular clearance required = γ and $h\colon \mathbb{R}^2 \to \mathbb{R}$, where $h(z)$ = known height of the container$-p_z$. Then,

$$\tan(\gamma) = \left(h(z) + r/\cos(\gamma)\right)/g\left(p_x, p_y, n_x, n_y\right) \tag{6.1}$$

$$\sin(\gamma)\, g\left(p_x, p_y, n_x, n_y\right) - \cos(\gamma)\left(h(z)\right) = r \tag{6.2}$$

Let $\tan(\alpha) = g\left(p_x, p_y, n_x, n_y\right)/h(z)$. Then,

$$\gamma = \arcsin\left(r/\sqrt{(h(z))^2 + g\left(p_x, p_y, n_x, n_y\right)^2}\right) + \alpha \tag{6.3}$$

Fig. 6.7 Statistical analysis for 50 trials of pick-up of pellets from the bin

The angular clearance offered as per current cylinder configuration $= \theta_{(n_x, n_y, n_z)} = \arctan\left(n_z / \sqrt{n_x^2 + n_y^2}\right)$.

The path is collision free if $\theta > \gamma$ else if $\theta < \gamma$ but $\gamma - \theta < 15°$ then the approach normal is adjusted to re-route the path and approach without collision. If the adjustment required is beyond 15 deg the suction gripper fails in pick-up thus those cases are skipped. Similar approach is employed for side of the cylinder. Choose the line of interest by shifting the cylinder axis sideways, by the radius of the cylinder. The above algorithm ensures the gripper rod walls do not collide with the container wall.

Steps are also taken to ensure the face of cylinder doesn't collide with the wall. This is done by ensuring that the minimum distance of any wall from the face centroid is greater than the projected radius of the face. The statistical analysis for the 50 trials done for pellets pickup are shown in Fig. 6.7 and are discussed in the next section.

6.2.4 Results and Discussion

We have implemented objects, i.e., pellets detection algorithm and the results for pose detection using data from both laser scanner and camera mounted on an industrial robot. At first, a sensor calibration methodology is used to enable the creation of cloud data in robot coordinate frame. Later a 3D model fitting based on local geometric features has been used. The data obtained from 3D point cloud was verified for further segmentation if required (in case of multiple detections) using the same region's image data. This helped in increasing the accuracy and speed of the processing. Finally, the detected pose is sent to the KUKA KR5 Arc controller.

With the number of trials conducted, the statistical analysis is presented. For the test procedure carried out, the time required for the pellet's pick-up from the bin and then place is approximately 17 s, and assuming 3 failures in 35 pick-up attempts, amortized cost accounting for unsuccessful is 22 s approx.

To discuss and show the results of the time taken by the robot to pick up, time to scan, time to place the pellet in an arranged manner is reported. Mean time to scan the bin and grab the image is 7 secs as shown in Fig. 6.8a. It is verified from the

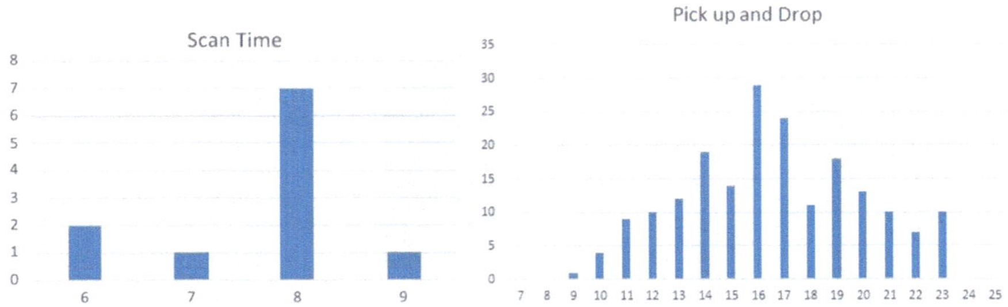

Fig. 6.8 Time for **a** scanning the bin and grab image, and **b** time to place/drop the pellet at a given position inside the workspace of the KUKA KR5 Arc

Table 6.1 Chart for pose identification, pick-up, and place

Scan time + Image grab time	7.6 s
Pick and place time	16.5 s
Segmentation time	350 ms
Pick-up success rate	90%

image that during experiment, different readings were taken for the scan time and as seen from Fig. 6.8, most of the trials were concentrated around 8 secs.

It is that the time taken for whole cycle, i.e., to pick up from the bin and transfer it to the desired position, is around 16 secs, moreover it can be seen from Fig. 6.8b that it is like a Gaussian curve, in which mean is about 16 and with the deviation of 2.5 was observed (Table 6.1).

6.3 Intermediate Steps

After the pellets were picked from the bin, they are either in horizontal or vertical orientations. This section describes the intermediate steps taken before the insertion of these pellets, namely, stacking of pellets, detection of hole, and approaching the tube (Fig. 6.10).

6.3.1 Stacking of Pellets

The intermediate step after picking the pelleted from the bin which are either in sleeping, i.e., horizontal position or vertical position, is for stacking them accordingly. Figure 6.9a shows the stacking of the pellet which is picked using the suction gripper

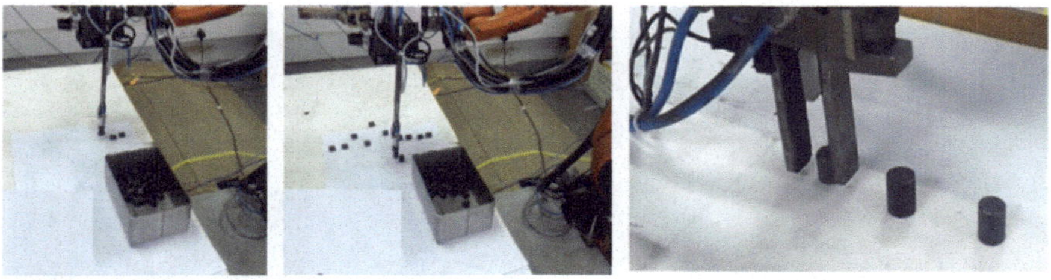

(a) (b) (c)

Fig. 6.9 Placement and pick-up of pellets

from the bin at the intermediate position. Similarly, the pellet picked in the vertical orientation is stacked as shown in Fig. 6.9b. The step is essential to change the gripper from suction to two-finger gripper for the insertion task. Since the suction gripper is made from the soft material and is flexible, it will not result in insertion of the pellet inside the tube with very less clearance. Figure 6.9c shows the array of the vertically placed pellet which is then picked using the two-finger grippers. For the sleeping pellets, the reconfiguration steps are shown in Fig. 6.10a–d. In an industrial setting, a simple example would be placing multiple toys in different boxes all kept on the assembly line. A rectangular grid will be defined for this purpose. Every time robot picks up a pellet, it goes to the next grid position and places the pellet in upright fashion. Figure 6.9c shows grid-making process. In an industrial setting, there might be requirement to assemble objects in other fashions rather than what was discussed earlier. It is commonly required that after picking up the desired object, robot may have to load a machine with these objects with no human intervention. This section discusses the method to load machines with such objects using computer program.

6.3.2 Detection of Hole

The image of the machine is processed during an edge detection algorithm, e.g., Canny edge detection [2] algorithm. The candidate centers are then selected from those points in this (two-dimensional) accumulator that are both above some given threshold and larger than all of their immediate neighbors. These candidate centers are sorted in descending order of their accumulator values, so that the centers with the most supporting pixels appear first. Next, for each center, all of the non-zero pixels (recall that this list was built earlier) are considered. These pixels are sorted according to their distance from the center. Working out from the smallest distances to the maximum radius, a single radius is selected that is best supported by the non-zero pixels. A center is kept if it has sufficient support from the non-zero pixels in the edge image and if it is a sufficient distance from any previously selected center. Figure 6.11 shows the image of different circles detected using Hough transform.

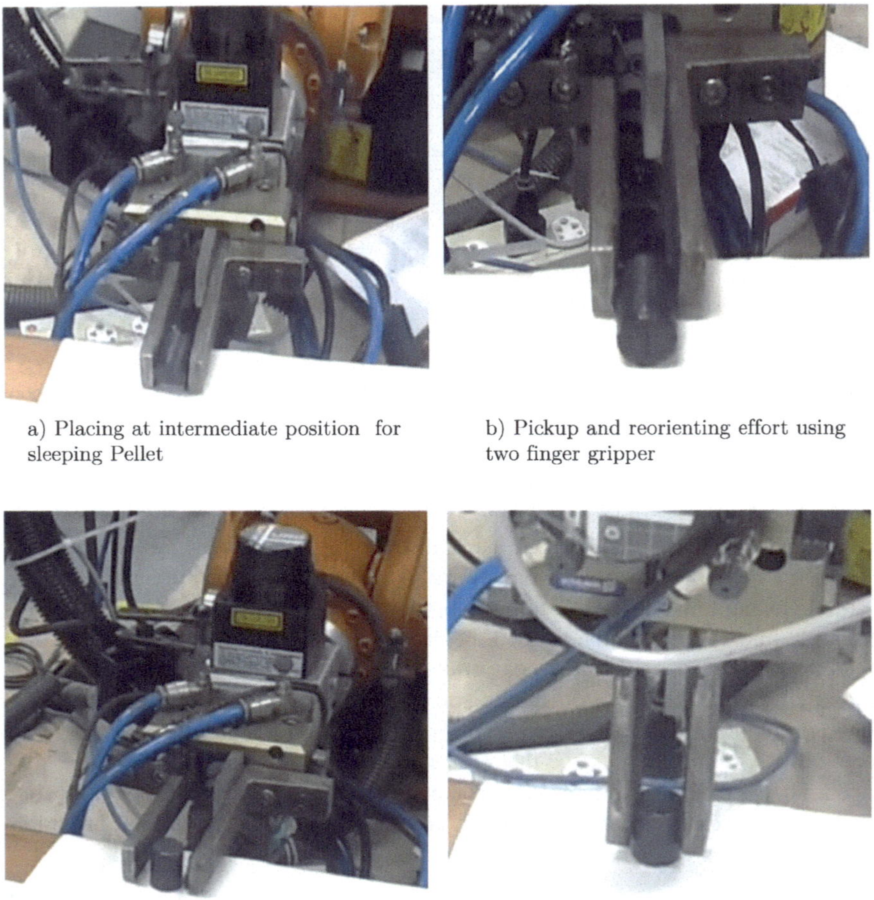

a) Placing at intermediate position for sleeping Pellet

b) Pickup and reorienting effort using two finger gripper

c) Reorienting pellet using gripper in upright position

d) Approach of gripper at upright position of pellet

Fig. 6.10 Pellet pick-up and reorienting to grip with two-finger gripper

6.3.3 Approaching Near the Tube

After getting the information about the center and radius of holes with reference to image coordinate system, we found the position of that hole in world coordinate system. As discussed in Chap. 5, one cannot find 3D world coordinate using single image without knowing any one of X-, Y-, or Z-coordinate. Hough circle detection algorithm gave centers and radii of different circles detected in given scene in image coordinate system and the laser scanner mounted gave the accurate level of hole detected. We knew the Z-coordinate of the hole and using this estimate the 3D world coordinate was estimated. To restrict insertion of pellet into a false input receptor

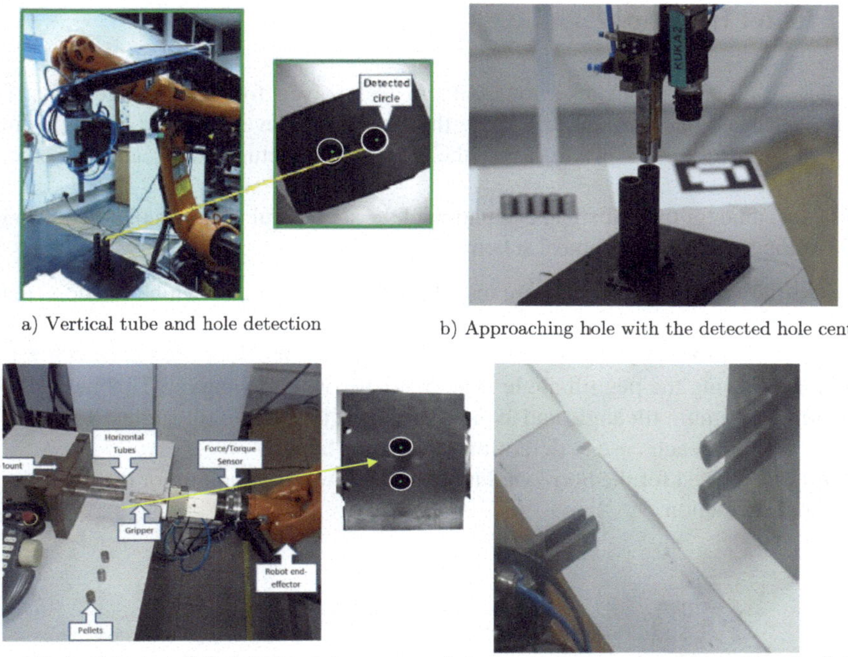

a) Vertical tube and hole detection

b) Approaching hole with the detected hole center

c) Horizontal tube and the hole detected

d) Approach of gripper at upright position of pellet

Fig. 6.11 Orienting the two-finger gripper holding the pellet near the tube assembly in vertical and horizontal cases

hole of the machine, we have filtered our list of detected circles by their radii in real world. Since we knew the center and radius of every circle in the image of the machine, we can compute point on circle having Y-coordinate same as center but with an increment in X-coordinate by value of radius. This was done in image coordinate system. After getting the pixel values, we re-computed its 3D coordinates. Using these algorithms and computer vision theories, we got a point on circle and center of the circle in world coordinate system. One can leverage this knowledge to easily find out the radius of the circle and filter out false circles by applying a threshold. The tubes where assembly is to be accomplished are precision manufactured. Thus, such a strategy will help us to detect the correct position of insertion in presence of false circle detections.

The detected hole in the vertical and horizontal tube positions is shown in Fig. 6.11a, c, respectively. The orientation of the robot's end effector approaching the tube in case of vertical and horizontal tubes is depicted in Fig. 6.11b, d, respectively. If the clearance between the tube and pellet is large Hough circle detection may be used to directly insert the pellet in tube. When clearance is less, the pellet will be brought close to the tube and then minimum depth search algorithm based on the active/passive control scheme as discussed in Chap. 5 is utilized. The insertion experimental results are discussed next.

6.4 Pellet Insertion

The peg-in-tube or pellet insertion problem was solved for two arrangements: a vertical case and a horizontal case using the algorithms developed and discussed in Chap. 5. This section will discuss about the experimental setup, results, and outcome.

The test plan for the statistical evaluation of the successful insertion of pellet using the developed vision and control scheme is listed below:

- Five different peg diameters were tested for the insertion task which are depicted in Fig. 6.12a.
- The test plan for vertical peg-in-tube insertion is that the depth-based search was tried by varying the peg tilt angle from 5° to 30° with 5° interval. This helped in deciding optimum tilt angle and avoiding cases of unsuitable tilt angles. Ten trials were taken for each case to determine success or failure.
- Similarly, the test for the horizontal insertion task was carried out for the optimum tilt angle selection.

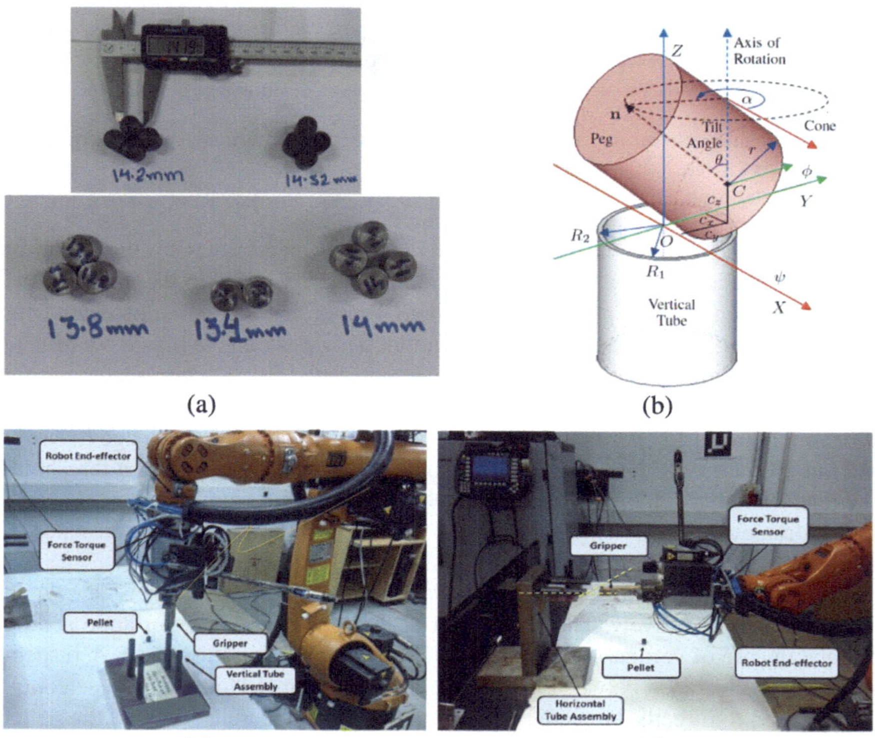

Fig. 6.12 Different pellets size, tilt angles, and the vertical and horizontal tube assembly

6.4.1 *Vertical Insertion*

The setup for the vertical peg-in-tube problem is seen in Fig. 6.12c. A set of such tubes were attached to a base mount as seen in the figure. A vertical tube setup was fabricated wherein inner diameter of the tube was 14 mm. The peg-in-tube assembly is accomplished using a combination of adaptive force control for manipulation of the robot on and around the tube and a depth-based localization algorithm for the detection of the position of the hole. Both of these are elucidated in the following sections. The force maintaining algorithm is a quintessential part of the hole localization method. This ensures safe interaction between the robot and the physical environment. Additionally, it helps in measuring the depth of the tool center point which will be used for determining the exact hole location. Force-maintaining algorithm for the peg in a hole insertion task was improved for steady force maintenance during search process of the hole after multiple trials after multiple simulation and physical experiments. The system uses a proportional control with position feedback and an external force control loop as suggested by [3]. The force control algorithm for orientation control was also implemented.

6.4.1.1 Test Results

The different peg sizes that were tested for the insertion task are depicted in Fig. 6.12a. Results of the above experiments that were performed will be discussed here. The tilt angle was varied from 5° to 25° with a 5° interval and a tilt angle of 30° was rejected due to an extremely poor performance. Tables 6.2, 6.3, 6.4, 6.5, and 6.6 depict the performances when the tilt angle was varied from 5° to 25°.

Table 6.2 Vertical peg-in-tube for tilt angle = 5°

S. No.	Time taken (in s) *	1st attempt	Other than 1st attempt
1	49	Yes	
2	84	–	2nd attempt
3	122	–	3rd attempt
4	48	Yes	
5	49	Yes	
6	84	–	2nd attempt
7	115	–	3rd attempt
8	82	–	2nd attempt
9	82	–	2nd attempt
10	84	–	2nd attempt
Mean	79.9	In 1st attempt 30% success	

Table 6.3 Vertical peg-in-tube for tilt angle = 10°

S. No.	Time taken (in s) *	1st attempt	Other than 1st attempt
1	46	Yes	
2	79	–	2nd attempt
3	49	Yes	
4	45	Yes	
5	47	Yes	
6	82	–	2nd attempt
7	84	–	2nd attempt
8	83	–	3rd attempt
9	45	Yes	
10	85	–	2nd attempt
Mean	64.5	In 1st attempt 50% success	

Table 6.4 Vertical peg-in-tube for tilt angle = 15°

S. No.	Time taken (in s) *	1st attempt	Other than 1st attempt
1	48	Yes	
2	46	Yes	
3	81	–	2nd attempt
4	47	Yes	
5	49	Yes	
6	51	Yes	
7	50	Yes	
8	83	–	2nd attempt
9	48	Yes	
10	81	–	2nd attempt
Mean	62.1	In 1st attempt 70% success	

Note that the unsuccessful first attempt while inserting the peg inside the tube is continued by the consecutive search of the hole based on the depth search algorithm for the second and third time. Hence, the time corresponding to the second and third attempts are having higher magnitude. The best performance, i.e., with minimum number of second or third attempt was obtained with 15 deg cone angle. Hence, 15° tilt angle was used for further experiments with different clearance.

After extensive experimentation, a tilt angle of 15° was used for the best results in the detection of the direction of the hole and the results have been shown here for that. Another set of experiments were performed for various peg-tube clearances ranging from 1 mm to 0.05 mm. The time of assembly for different clearances is depicted in Fig. 6.13.

Table 6.5 Vertical peg-in-tube for tilt angle = 20°

S. No.	Time taken (in s) *	1st attempt	Other than 1st attempt
1	84	–	2nd attempt
2	79	–	2nd attempt
3	52	Yes	
4	83	–	2nd attempt
5	50	Yes	
6	53	Yes	
7	115	–	3rd attempt
8	114	–	3rd attempt
9	82	–	2nd attempt
10	52	Yes	
Mean	76.3	In 1st attempt 40% success	

Table 6.6 Vertical peg-in-tube for tilt angle = 25°

S. No.	Time taken (in s) *	1st attempt	Other than 1st attempt
1	55	Yes	
2	54	Yes	
3	85	–	2nd attempt
4	84	–	2nd attempt
5	81	–	2nd attempt
6	110	–	3rd attempt
7	50	Yes	
8	51	Yes	
9	84	–	2nd attempt
10	121	–	3rd attempt
Mean	77.5	In 1st attempt 40% success	

6.4.2 Horizontal Insertion

The setup for the horizontal peg-in-tube assembly is shown in Fig. 6.12d and is similar to the vertical setup. The dimensions of the tube are the same as the ones discussed earlier. One pair of the tubes have been attached to a base mount as shown in Fig. 6.12. The peg-in-tube assembly is accomplished using the same combination of adaptive force control for manipulation of the robot on and around the tube and a depth-based localization algorithm (modified for the horizontal setup) for the detection of position of the hole. Both of these are elucidated in the following sections.

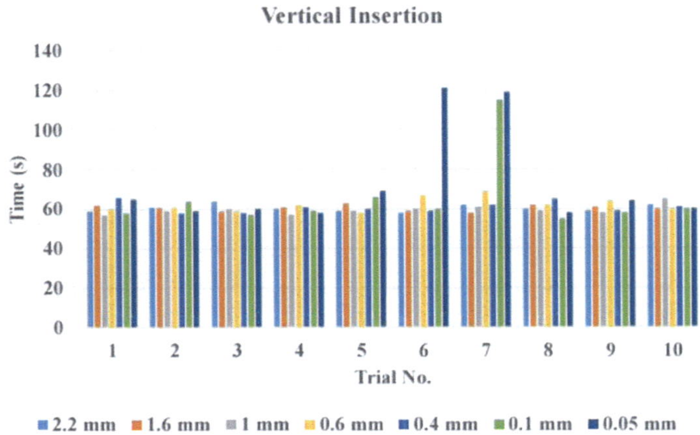

Fig. 6.13 Vertical insertion data for various clearances and tilt angle = 15°

6.4.2.1 Depth-Based Localization for Horizontal Insertion

The horizontal peg-in-tube is different from the vertical case where the pellets fall with the effect of gravity inside the tube. Here the pellets need to be inserted by a push inside the horizontal tube. A setup has been made by modifying the gripper with a pneumatic supply. Compressed air is used to insert the pellet inside the tube once pellet is aligned with the center of the tube. Similar to the vertical case, once the peg came close to the tube, the peg is tilted and made to approach the tube till contact is made. Then the peg is rolled over the hole while it is tilted at the fixed angle. A constant reaction force of 10N was maintained to ensure continuous contact with the surface. Once the contact is established the peg was made to roll over its edge in a conical path. The inclination from the horizontal in two orthogonal directions is given below:

$$A = \sin^{-1}(\cos\alpha \sin\theta)\,;\, C = 90 + \sin^{-1}(\sin\alpha \sin\theta) \tag{6.4}$$

where A and C are inclination of peg axis from the vertical Y axis along XY- and YZ-planes, respectively. The half-angle of the cone apex is θ and α is the angle of current tilting plane from the start tilt plane.

The system keeps monitoring the horizontal movement of the peg along the Y-axis. The hole direction is given by the least movement above the hole surface and can be calculated as

$$\triangle v_x = k\cos(\alpha_{\text{minimum}}) \text{ and } \triangle v_z = k\sin(\alpha_{\text{minimum}}) \tag{6.5}$$

where $\triangle v_x$ and $\triangle v_z$ are the velocity of the peg along X- and Z-directions, respectively, and k is a constant. The angle $\alpha_{minimum}$ is the α at minimum depth observed during the rotation cycle.

Once the hole direction is calculated, the peg is moved till it finds a significant drop in normal force perpendicular to the surface (this is 5N in the current case).

Steps for the algorithm
Approach tilted till contact is established
Store Y_minimum and maintain a constant force (10N)
for $\alpha = 0$ to 360 degrees
{
If Y_current < or = Y_minimum
{
Y_minimum = Y_current
α_minimum = α_current
}
}
Move towards hole direction
$\triangle x = 0.03 \times \cos(\alpha\_minimum)$
$\triangle z = 0.03 \times \sin(\alpha\_minimum)$

Drop the pellet if Force_current along Y is less than 4N OR Y< Y_contact in flat condition.

The above experiments were performed and the results are listed in tabular form. The tilt angle was varied from 5° to 25° with a 5° interval and a tilt angle of 30° was rejected due to an extremely poor performance as was the case with the vertical insertion. Tables 6.7, 6.8, 6.9, 6.10, and 6.11 represent the results of these experiments.

Table 6.7 Horizontal peg-in-tube for tilt angle = 5°

S. No.	Time taken (in s) *	1st attempt	Other than 1st attempt
1	75	–	2nd attempt
2	73	–	2nd attempt
3	45	Yes	
4	76	–	2nd attempt
5	49	Yes	
6	41	Yes	
7	77	–	2nd attempt
8	110	–	3rd attempt
9	112	–	3rd attempt
10	61	–	2nd attempt
Mean	71.9	In 1st attempt 30% success	

Table 6.8 Horizontal peg-in-tube for tilt angle = 10°

S. No.	Time taken (in s) *	1st attempt	Other than 1st attempt
1	49	Yes	
2	52	Yes	
3	80	–	2nd attempt
4	85	–	2nd attempt
5	79	–	2nd attempt
6	52	Yes	
7	54	Yes	
8	120	–	3rd attempt
9	54	Yes	
10	117	–	3rd attempt
Mean	74.2	In 1st attempt 50% success	

Table 6.9 Horizontal peg-in-tube for tilt angle = 15°

S. No.	Time Taken (in s) *	1st attempt	Other than 1st attempt
1	52	Yes	
2	55	Yes	
3	54	Yes	
4	82	–	2nd attempt
5	82	–	2nd attempt
6	55	Yes	
7	54	Yes	
8	90	Yes	
9	91	–	2nd attempt
10	57	Yes	
Mean	67.2	In 1st attempt 70% success	

Similar to the vertical peg insertion, the success rate of 70% at a tilt angle of 15° is exhibited if we consider only the insertion during the first attempt as a success. However, if we also consider the insertion during the second attempt as a success, the success rate would jump up to around 95% as is the case with vertical insertion. As attempted earlier, the assembly task has been attempted with variations in the clearance between peg and the tube. The clearance was varied from 2.2 mm to 0.05 mm. The time taken for different attempts is shown in Fig. 6.14. Once the insertion is done successfully for the first time, the coordinates are saved for the location of the hole center, and the next insertion can be performed in much lesser time that will include only the pick and insertion path. The algorithm for the assembly task of pellets inside the tube was successfully implemented for the two orientations of the tube, i.e., one when it is parallel to the direction of the gravity and another in

Table 6.10 Horizontal peg-in-tube for tilt angle = 20°

S. No.	Time taken (in s) *	1st attempt	Other than 1st attempt
1	53	Yes	
2	55	Yes	
3	85	–	2nd attempt
4	84	–	2nd attempt
5	86	–	2nd attempt
6	56	Yes	
7	57	Yes	
8	120	–	3rd attempt
9	116	–	3rd attempt
10	86	–	2nd attempt
Mean	80.1	In 1st attempt 40% success	

Table 6.11 Horizontal peg-in-tube for tilt angle = 25°

S. No.	Time taken (in s) *	1st attempt	Other than 1st attempt
1	58	Yes	
2	85	–	2nd attempt
3	84	–	
4	121	–	3rd attempt
5	55	Yes	
6	59	Yes	
7	58	Yes	
8	122	–	3rd attempt
9	119	–	3rd attempt
10	89	–	2nd attempt
Mean	85.0	In 1st attempt 40% success	

which it is kept horizontal to the gravity. In vertical case, the pellets once aligned get inserted inside the tube with the gravitational force. While in the case of horizontal assembly, an external pneumatic push was devised to insert the pellet.

6.5 Summary

In this chapter, the experimental evaluation of the algorithms developed for the assembly task was presented. The two extreme cases for the assembly task, i.e., insertion of pellets in a vertical tube and then the horizontal tube were discussed. The vision-based pick-up of pellets from the bin shows the success rate of picking as high as 90%

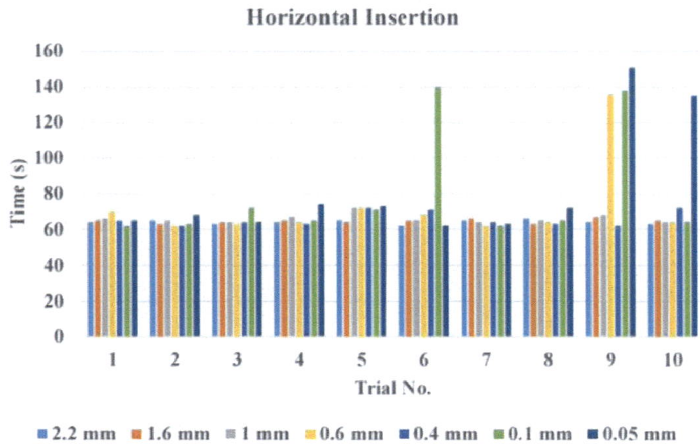

Fig. 6.14 Vertical insertion data for various clearances and tilt angle = 15°

for 50 runs. Moreover, the optimum tilt angle found for the minimum depth-based search algorithm was 15°, at which the successful insertion in the first attempt was 70%. The integration approach presented is generic and can be adopted to evaluate the optimum tilt angle for the minimum depth-based search approach and the success rate evaluation for different sizes of the peg-in-tube assembly task. The validation of the vision-based algorithms and the minimum depth-based search algorithms with active–passive force control are validated. This approach can be used for any high-precision assembly task to estimate the proper parameters as per the setup used.

References

1. Chittawadigi, R.G., Hayat, A.A., Saha, S.K.: Robust measurement technique for the identification of DH parameters of serial robots. In: The 3rd IFToMM International Symposium on Robotics and Mechatronics, pp. 1–6. ISRM (2012)
2. Canny, J.: A computational approach to edge detection. IEEE Trans. Pattern Anal. Mach. Intell. **8**(6), 679–698 (1986)
3. Winkler, A., Suchy, J.: Position feedback in force control of industrial manipulators—an experimental comparison with basic algorithms. In: IEEE International Symposium on Robotic and Sensors Environments (ROSE 2012), pp. 31–36 (2012)

Chapter 7
Conclusion

Manufacturing industries use input raw materials and add value to them through customized procedures. Material handling and organization is a principal part of this process. However, there is constant evolution in how manufacturing industries operate driven by profit, productivity, process reliability, etc. The fact evidences this at the beginning of the twentieth century industrial robots made a foray and started replacing manual labor in some aspects like painting and welding operations. By the beginning of the twenty-first century, the focus shifted to operating robots in unstructured environments. However, a significant impediment to the proliferation of these techniques is the lack of reliable visual and force feedback and suitable fusion of these multiple modes of data to allow the robot to perceive and adapt to the external environment. This book presents two examples of this process, i.e., integration of bin picking and subsequent assembly to illustrate how an industrial robot can be adapted to perceive the external environment and perform tasks intelligibly.

The broad areas focused in this monograph are identification of pose for robotic pick-up, subsequent localization for assembly, and force control using identified dynamic model for safer assembly. They are illustrated with a cylindrical object and its insertion inside a hole similar to that commonly encounters in manufacturing industries.

Pose estimation using the fusion of information from robot-mounted sensors for automated picking using an industrial robot was discussed first. This approach is suitable for automating the assembly task in dangerous environments like nuclear and chemical industries. In such situations, conventional feature extraction techniques and segmentation techniques cannot be applied, given the uniqueness of the conditions. Pose estimation of featureless components commonly encountered in manufacturing environments using RGB camera and a 2D laser sensor has been presented in this monograph. Given the unique surface properties of the object, i.e., cylindrical shape, with dark color which absorbs incident illumination, conventional

A. A. Hayat et al., *Vision Based Identification and Force Control of Industrial Robots*, Studies in Systems, Decision and Control 404,
https://doi.org/10.1007/978-981-16-6990-3_7

solutions using off-the-shelf RGB-D sensors like Kinect or Realsense camera could not be applied. Therefore, a dedicated 2D sensor capable of position measurement in such adverse situation was used and the necessary step was the extrinsic calibration of the sensors. After this, it was possible to obtain the position information in the robot coordinate frame and, most importantly, 2D information was converted to point cloud data for the region of interest. Point Cloud Data (PCD) is grabbed through scanning the region by moving the robot-mounted sensor over the region of interest. Point cloud data grabbed in this fashion could be combined with the RGB information from a camera mounted on the robot end effector using the hand–eye relation. The local information from the PCD could be used to fit object model to enable estimation of the pose. Given the extrinsic transformation between the end-effector mounted sensors and the robot, it was possible to project the point cloud to the image data to detect disturbance during the picking up operation. This could avoid time-consuming and resource-intensive area scans in a repetitive fashion, thus speeding up the process. Considering the practical aspects like collision detection, a working model of the system was built and the experimental results are presented. The complete system has been demonstrated in videos which are available as a part of this monograph.

To better understand the system's performance, a detailed sensitivity analysis of the proposed method has also been presented in Chap. 3. The picking up operation involves the robot kinematics, extrinsic calibration of the sensors, and intrinsic calibration of the camera, and each of these sources can contribute to errors in pose estimation and final picking up of the object. The errors from each source, i.e., kinematics, extrinsic, and intrinsic calibration, have been studied in isolation. Sobel's method was used to model the effect of individual parameters on the end-effector pose as well. Finally, the combined effect of all these factors has been quantified to provide bounds on the expected errors in pose estimation and pick-up.

Another crucial step in manufacturing industries is the assembly of different items to produce the end product. The process is a culmination of sensory information from multiple sources which is a tough problem to automate. Multiple sensory information and their fusion are difficult given the conditions like tight tolerance, possibility of damage due to errors in positioning, etc. An application of assembly in tight tolerance was discussed. An industrial robot needs to act in a compliant manner during the assembly, and this requires operation in gravity compensation and force control mode. The system parameters, namely, the kinematic and dynamic parameters, need to be identified to enable this. Kinematic identification using a geometrical method utilizing only the data from the vision sensor mounted on it for the picking operation is presented. The dynamic parameters were identified using an inverse dynamics model based on DeNOC (decoupled natural orthogonal complement)-based formulation. The data from the robot controller was used to identify the base inertial parameters. The performance of identification was evaluated using test trajectory data and the results were discussed. The identified system model could enable active–passive compliance using joint-level torque measurement and externally mounted force/torque sensor. This allowed operating the robot in force control for the assembly operation. However, given the performance measurement of the robot, the assembly in small clearances was beyond the operational capability of the robot. Therefore, a rough localization using vision data was refined using a tilt-rotation search method based on the feedback from the force–torque sensor mounted before the end effector. The proposed method could be used to localize the peg even in conditions where the clearance between the peg and tube is lesser than the position repeatability and accuracy of the robot. The experimental results in the stated conditions have been presented. A conventional industrial robot cannot accomplish assembly of peg in a tube under such tight tolerances in normal conditions.

One unique aspect of the monograph is that all the proposed methods have been practically implemented on an industrial robot, and results are discussed. Videos of the processes also accompany the explanations. Therefore, the contents of this monograph will surely be of interest to researchers as well as industrial practitioners. Apart from a unilateral focus on either bin-picking, assembly in already existing works, a complete solution to both the processes and the practical requirements in the hardware is presented. The uniqueness of this treatment is that the discussion caters to the requirements of both the practitioners driven by the motive of building a functional/working model and the academician who will try to model the system in further detail, like sensitivity analysis. The authors hope that the treatment will benefit the overall robotics community with a balance of practical aspects and theory. It is surely a worthwhile starting point to further research and refine these processes by incorporating techniques like artificial intelligence.

7.1 Limitations and Future Directions

Some limitations of the methods presented in this book are presented here. It deals specifically with picking up and assembling cylindrical objects using conventional techniques considering the possibility of control over all aspects of the process. Though kinematic identification using vision sensor was proposed, some parameters could not be determined accurately. Sensitivity analysis mainly considered the robot forward kinematics, i.e., the change in pose given the variation of the parameters. However, the inverse relation is also relevant while considering kinematic identification. Dynamic identification was illustrated mainly using robot's data, without external measurements of the pose, force, etc. Finally, localization did not use continuous feedback from visual and force/torque data or their fusion. Considering these aspects, some future directions of research can be envisioned as follows:

- Pose measurement using deep learning techniques based on point cloud and RGB data.
- A second-level system identification using learning techniques to enable better fit between predicted and achieved results.
- Fusion of vision and force/torque data for assembly to increase the speed of assembly.

- Adapt the method of assembly using state-of-the-art robots like KUKA iiwa which has inherent force/control strategy.
- Learning-based identification of kinematic and dynamic parameters.
- Data-driven autonomy of the assembly task including different layers of safety incorporated.
- Use of a combination of serial and parallel manipulators to enable rough and fine localization for assembly.
- Comparison of the assembly task of the featureless objects using the state-of-the art robot that comes with embedded haptic sensors in the end-effector and joint-torque sensors.

Please scan above for videos and refer Appendix C.3 for files detail

Appendix A
Vision and Uncertainty Analysis

In this appendix, several basics of vision systems and uncertainty are explained.

A.1 Image Formation

Mathematical expressions governing image formation in a machine vision system involve two types of parameters, i.e., internal and external parameters. Internal parameters consist of effective focal length f, lens distortion coefficients (k_1 to k_6 and p_1 to p_2), and image coordinate of principal point ($\mathbf{c}$). External parameters consist of rotation ($\mathbf{Q}$) and translation ($\mathbf{t}$) parameters. As in the case of robots, the estimation of these parameters is termed as calibration. Calibration process involves placing a pattern of known size and processing its image (Fig. A.1) to estimate the camera parameters. Once these parameters are known, the image coordinate information can be converted to real-world dimensions. Calibration is done utilizing the basic phenomenon of image formation in a camera. In Fig. A.2, we consider a point with coordinates $\mathbf{x} \equiv [x, y, z, 1]^{\mathsf{T}}$ in world coordinate system (represented with W). A ray of light passes from this point through the lens system (represented by camera coordinate system C) and is focused on the sensor plane (represented by image coordinate system I) to obtain an image with pixel coordinates $\mathbf{u} \equiv [u, v, 1]^{\mathsf{T}}$. It is represented in homogenous coordinate system as $\mathbf{p} \equiv s\mathbf{u}$. Scaling of image is represented by s. Lens optical axis intersects with the camera sensor at the image coordinates given by $\mathbf{c} \equiv [c_x, c_y]^{\mathsf{T}}$. This is called principal point. The location of the origin of coordinate frame C in terms of coordinate frame W is $\mathbf{t} \equiv [t_x, t_y, t_z]^{\mathsf{T}}$. The corresponding rotation matrix is given by $\mathbf{Q}$ (may be represented in terms of roll, pitch, and yaw angles). The camera extrinsic matrix $\mathbf{T} \equiv [\mathbf{Q}|\mathbf{t}]$ of size 3×4 is obtained by utilizing these. The camera constant f is related to the focal length and the pixel-wise resolution of the image. Coordinate transformation and projective transformation are represented by

A. A. Hayat et al., *Vision Based Identification and Force Control of Industrial Robots*, Studies in Systems, Decision and Control 404,
https://doi.org/10.1007/978-981-16-6990-3

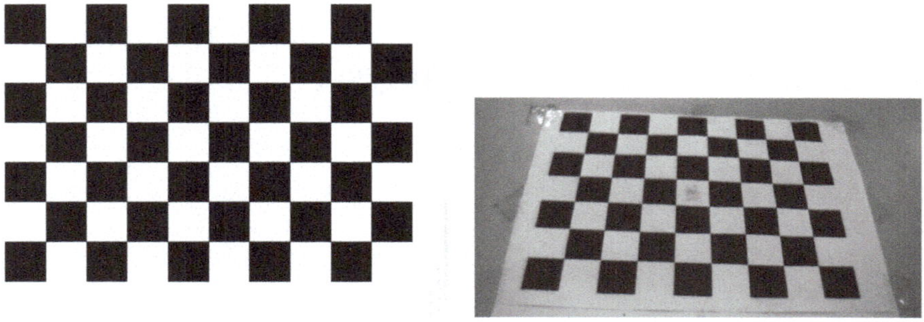

(a) (b)

Fig. A.1 Calibration grid

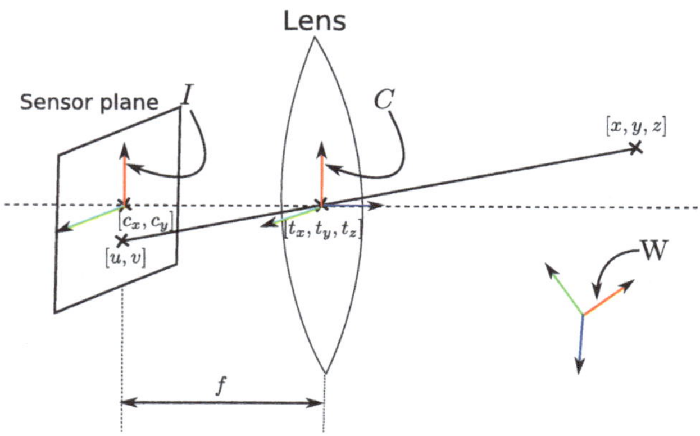

Fig. A.2 Image formation

$$s\mathbf{u} = \mathbf{CTx} \tag{A.1}$$

where the camera intrinsic matrix **C** is given by

$$\mathbf{C} \equiv \begin{bmatrix} f_x & 0 & c_x \\ 0 & f_y & c_y \\ 0 & 0 & 1 \end{bmatrix}$$

The factors f_x and f_y are effective focal length in X-axis and Y-axis of the image.

The images normally suffer from distortion which is mathematically depicted in the following manner:

$$\mathbf{u} = \mathbf{C}\bar{\mathbf{u}} \tag{A.2a}$$

where

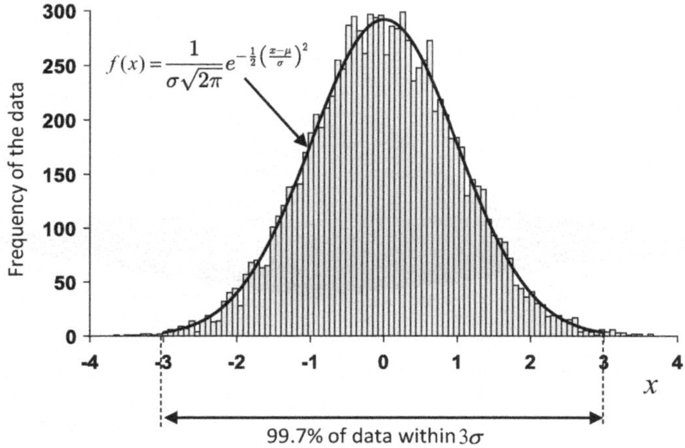

Fig. A.3 Normal (or Gaussian) distribution

$$\bar{\mathbf{u}} \equiv \begin{bmatrix} x^u \frac{1+k_1\bar{r}+k_2\bar{r}^2+k_3\bar{r}^3}{1+k_4\bar{r}+k_5\bar{r}^2+k_6\bar{r}^3} + 2p_1x^uy^u + p_2(\bar{r}+2x^{u^2}) \\ y^u \frac{1+k_1\bar{r}+k_2\bar{r}^2+k_3\bar{r}^3}{1+k_4\bar{r}+k_5\bar{r}^2+k_6\bar{r}^3} + p_1(\bar{r}+2y^{u^2}) + 2p_2x^uy^u \\ 1 \end{bmatrix} \tag{A.2b}$$

It may be noted that the radial distortion parameters are represented by k_1, k_2, k_3, k_4, k_5, and k_6, whereas tangential distortion parameters are represented by p_1 and p_2 [1]. The un-distorted coordinates are given by $\mathbf{x}^u \equiv [x^u, y^u, 1]^\mathsf{T}$, and the radial distance of the image coordinate is given by $\bar{r} \equiv x^{u^2} + y^{u^2}$. Perspective transformation $\mathbf{x}^u$ is obtained as

$$\mathbf{x}^u \equiv \frac{1}{z^c}\mathbf{T}\mathbf{x} \tag{A.2c}$$

Note that z^c is the Z-coordinate of the point $\mathbf{x}$ expressed in coordinate frame C.

A.2 Gaussian Distribution

Gaussian or Normal distribution is a bell curve-shaped distribution, as shown in Fig. A.3. Physical quantities often follow this distribution with independently drawn random samples. Figure A.3 was drawn with 10, 000 data points with mean as zero and standard deviations (SD), $\sigma = 1$. Figure A.3 shows the normal distribution of a single variable which is also denoted as $\mathcal{N}(0, \sigma^2)$ where $\mathcal{N}$ denotes normal distribution with zero mean and variance[1] σ^2.

---

[1] variance = (standard deviation)^2.

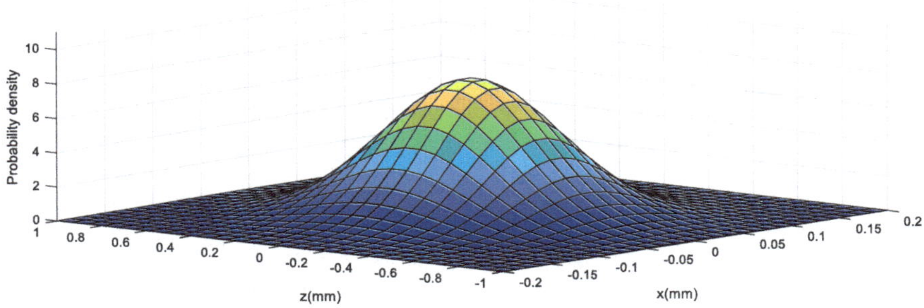

Fig. A.4 Probability density function

For multivariate distribution, i.e., say for vector **x** with n variables is denoted as $\mathcal{N}(\mathbf{0}, \mathbf{V})$, $\mathbf{0}$ is a zero vector of size $n \times 1$, and $\mathbf{V}$ is a covariance matrix as

$$Var(\mathbf{x}) = \begin{pmatrix} \sigma_{x_1}^2 & \sigma_{x_1x_2} & \cdots & \sigma_{x_1x_n} \\ \sigma_{x_1x_2} & \sigma_{x_2}^2 & \cdots & \sigma_{x_2x_n} \\ \vdots & \vdots & \ddots & \vdots \\ \sigma_{x_1x_n} & \sigma_{x_2x_n} & \cdots & \sigma_{x_n}^2 \end{pmatrix}$$
$$= \mathbf{V}$$

A variance-covariance (VCV) matrix is square, symmetric, and are always positive definite, i.e., all of the eigenvalues must be positive. For independent or unbiased variables, the VCV matrix will have the off-diagonal terms zero resulting in the covariance matrix $\mathbf{V} \equiv diag(\sigma_{x_1}^2, \cdots, \sigma_{x_n}^2)$. The distribution of the uncertainty in laser frame is depicted as Gaussian curve in Fig. A.4 for two variables.

Reference

1. OpenCV: Camera calibration and 3D reconstruction (2020). http://docs.opencv.org/2.4/modules/calib3d/doc/camera_calibration_and_3d_reconstruction.html. [Online accessed 06-October-20]

Appendix B
Robot Jacobian

The derivation of the Jacobian matrix using the forward kinematic transformation matrices for serial-chain robotic system is discussed here.

B.1 Jacobian

Referring to Fig. B.1, the $6 \times n$ Jacobian for a serial-chain robot manipulator is given by

$$\mathbf{J} = \begin{bmatrix} \mathbf{e}_1 & \mathbf{e}_2 & \cdots & \mathbf{e}_n \\ \mathbf{e}_1 \times \mathbf{a}_{1e} & \mathbf{e}_2 \times \mathbf{a}_{2e} & \cdots & \mathbf{e}_n \times \mathbf{a}_{ne} \end{bmatrix} \tag{B.1}$$

where $\mathbf{e}_i \equiv [e_{ix}\ e_{iy}\ e_{iz}]^T$ is the unit vector parallel to the axis of rotation of the ith joint represented in the inertial frame, and $\mathbf{a}_{ie}$ is the vector joining the ith joint frame to the end-effector frame as shown in Fig. B.1. The 4×4 forward kinematic homogeneous transformation matrix of the ith joint frame with respect to the robot's base $\left[\mathbf{T}_{i-1}\right]_1$ comprises the unit vector parallel to the axis of rotation of the link #i, i.e., $\mathbf{e}_i \equiv [e_{ix}\ e_{iy}\ e_{iz}]^T$ as its third column, and the position vector of the frame attached to the link #$(i-1)$, i.e., $\mathbf{o}_i \equiv [o_{ix}\ o_{iy}\ o_{iz}]^T$, as its fourth column, namely,

$$\left[\mathbf{T}_{i-1}\right]_1 = \begin{bmatrix} \cdot & \cdot & \mathbf{e}_i & \mathbf{o}_i \\ 0 & 0 & 0 & 1 \end{bmatrix} \tag{B.2}$$

where $i = 1, 2, \cdots, n$ are the joint or link numbers. Moreover, $\mathbf{e}_1 \equiv [0\ 0\ 1]$ and $\mathbf{o}_1 \equiv [0\ 0\ 0]$. The elements of the matrix $\left[\mathbf{T}_{i-1}\right]_1$ shown as "·" in (B.2) are not given as these are not required in (B.1). The ith frame is attached at the joint O_i connecting

A. A. Hayat et al., *Vision Based Identification and Force Control of Industrial Robots*, Studies in Systems, Decision and Control 404,
https://doi.org/10.1007/978-981-16-6990-3

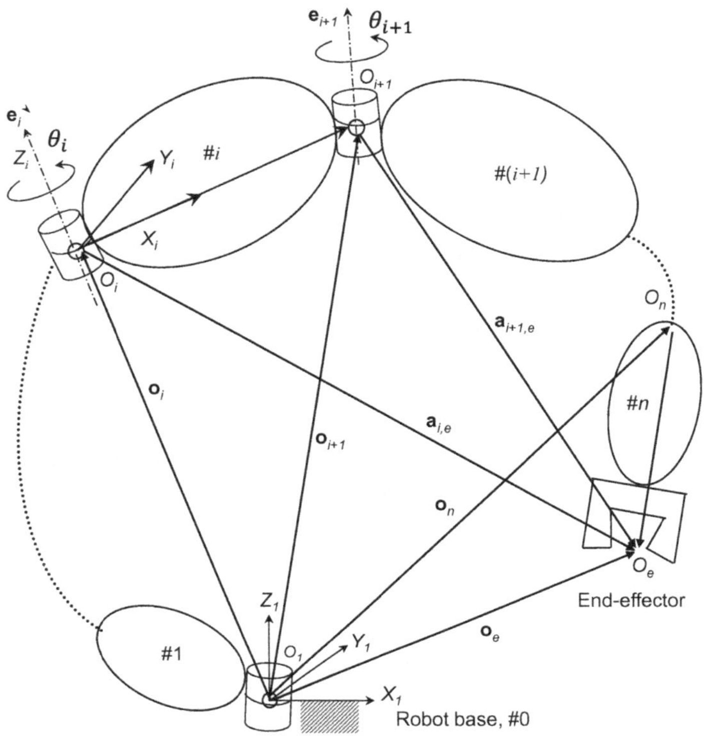

Fig. B.1 A serial-chain system with end effector

links #$(i-1)$ and #i. The vectors $\mathbf{e}_i$ and $\mathbf{o}_i$ are extracted from $\left[\mathbf{T}_{i-1}\right]_1$, whereas the vector $\mathbf{o}_e$, i.e., $\mathbf{o}_{n+1}$ is obtained from $[\mathbf{T}_n]_1$. Next, the vectors $\mathbf{a}_{ie}$ were calculated as

$$\mathbf{a}_{ie} = \mathbf{o}_e - \mathbf{o}_i \quad (B.3)$$

Thus, the Jacobian matrix $\mathbf{J}$ was evaluated using (B.2) and (B.3).

B.2 DH Parameters

In this appendix, the definitions of Denavit–Hartenberg (DH) parameters are presented for completeness of the paper. Note that, four parameters, namely, b_i, θ_i, a_i, and α_i [1], relate the transformation between two frames i and $i+1$ which are rigidly attached to two consecutive links #$(i-1)$ and #(i), respectively, as shown in Fig. B.2. Their notations and descriptions are summarized in Table B.1.

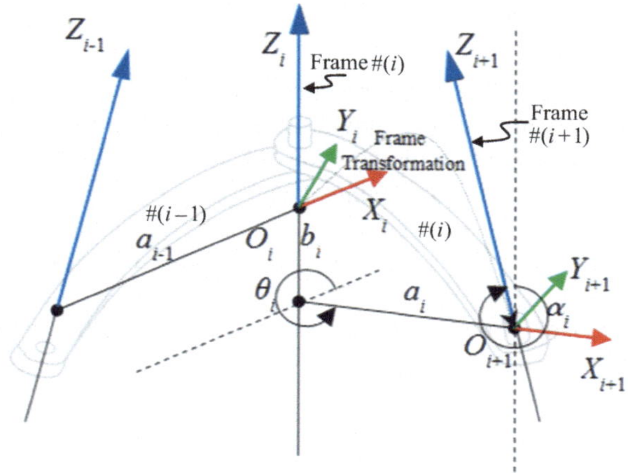

Fig. B.2 Links and DH parameters

Table B.1 Notations and descriptions of the DH parameters [1, 2]

Parameters (name)	Description[a]
b_i (Joint offset)	$X_i \xrightarrow[@Z_i]{\perp,\text{distance}} X_{i+1}$
θ_i (Joint angle)	$X_i \xrightarrow[@Z_i]{ccw,\ \text{rotation}} X_{i+1}$
a_i (Link length)	$Z_i \xrightarrow[@X_{i+1}]{\perp,\text{distance}} Z_{i+1}$
α_i (Twist angle)	$Z_i \xrightarrow[@X_{i+1}]{ccw,\ \text{rotation}} Z_{i+1}$

[a]In the table read symbol $\longrightarrow$ as "and", $\perp$ as "perpendicular", @ as "along"and ccw as "counterclockwise".

The guidelines followed in this paper to avoid redundancy of DH nomenclature (when two consecutive joints are parallel) are mainly:

- All the joint axis vectors (JAV) were identified first as per the formulation given in Chap. 4.
- The base or mounting plane of the robot was identified by fitting the points to plane.
- First frame with its origin was placed at the intersection of the first axis to the base plane.
- From the first origin, we moved to the next joint axis to place the origin at the intersection of two axes.
- In case of two parallel joint axis (joints 2, 3, and 5 are parallel in Fig. 4.11a), the origin obtained for joint axis 2 is used to place a perpendicular on the third axis, and so on.

B.3 Dual Vector

Figure B.3 shows the directions of two joint axes denoted with unit vectors $\mathbf{n}_i$ and $\mathbf{n}_{i+1}$ obtained using singular value decomposition (SVD) and the two vectors $\mathbf{c}_i$ and $\mathbf{c}_{i+1}$ representing the locations of the centers of the two circles, i.e., C_i and C_{i+1}. The dual vector $\hat{\mathbf{n}}_i$ is defined as

$$\hat{\mathbf{n}}_i = \mathbf{n}_i + \varepsilon(\mathbf{c}_i \times \mathbf{n}_i) = \mathbf{n}_i + \varepsilon \mathbf{n}_i^* \tag{B.4}$$

where $\mathbf{n}_i^* = (\mathbf{c}_i \times \mathbf{n}_i)$ and $\varepsilon^2 = 0$. Similarly, the dual vector associated to the $(i+1)$st joint is given by

$$\hat{\mathbf{n}}_{i+1} = \mathbf{n}_{i+1} + \varepsilon(\mathbf{c}_{i+1} \times \mathbf{n}_{i+1}) = \mathbf{n}_{i+1} + \varepsilon \mathbf{n}_{i+1}^* \tag{B.5}$$

The dual unit vector along $\mathbf{n}_i$, denoted by $\hat{\mathbf{k}}_i$, is then calculated as

$$\hat{\mathbf{k}}_i = \frac{\hat{\mathbf{n}}_i}{|\hat{\mathbf{n}}_i|} = \mathbf{n}_i + \varepsilon(\mathbf{n}_i \times \mathbf{n}_i^*) \times \mathbf{n}_i = \mathbf{n}_i + \varepsilon \mathbf{k}_i^* \tag{B.6}$$

where

$$\mathbf{k}_i = \mathbf{n}_i \ \text{ and } \ \mathbf{k}_i^* = (\mathbf{n}_i \times \mathbf{n}_i^*) \times \mathbf{n}_i \tag{B.7}$$

Similarly, the dual unit vector $\hat{\mathbf{k}}_{i+1}$ is obtained as

$$\hat{\mathbf{k}}_{i+1} = \mathbf{k}_{i+1} + \varepsilon \mathbf{k}_{i+1}^*, \quad \text{where } \mathbf{k}_{i+1} \equiv \mathbf{n}_{i+1} \tag{B.8}$$

Using definition of the DH parameters given in Saha (2014b), the joint axis is defined along the Z_i-axis whereas the common normal between Z_i and Z_{i+1} defines the

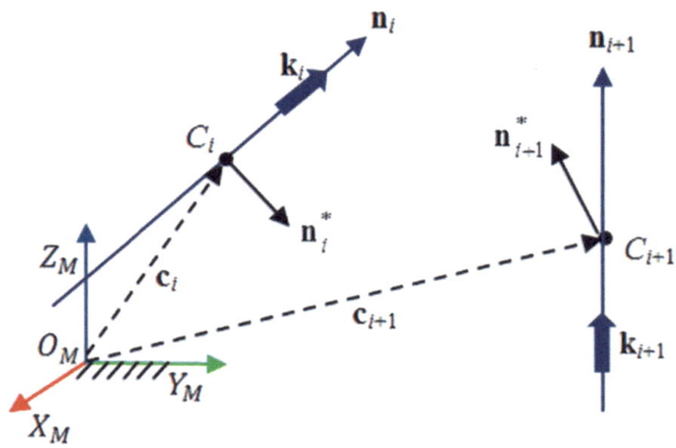

Fig. B.3 Joint axis and a point on it to represent a dual number

direction for X_{i+1}-axis. The Y_{i+1}-axis is then defined as per the right-hand screw rule which allows us to find the dual unit vector along X_{i+1}-direction, i.e., $\hat{\mathbf{i}}_{i+1}$ as

$$\hat{\mathbf{i}}_{i+1} = \hat{\mathbf{n}}_i \times \hat{\mathbf{k}}_{i+1} \tag{B.9}$$

Following the definition of a dual angle between two dual unit vectors, as designed in [3], denoted as $\hat{\theta}_i = \theta_i + \varepsilon b_i$ or $\hat{\alpha}_i = \alpha_i + \varepsilon a_i$, depending on the coordinate frame. By applying dot[2] and cross[3] products of two sets of dual unit vectors, one gets the following:

$$cos\hat{\alpha}_i = \hat{\mathbf{k}}_i \cdot \hat{\mathbf{k}}_{i+1} \text{ and } \cos\hat{\theta}_i = \hat{\mathbf{i}}_i \cdot \hat{\mathbf{i}}_{i+1} \tag{B.10}$$

$$\sin\hat{\alpha}_i = (\hat{\mathbf{k}}_i \times \hat{\mathbf{k}}_{i+1}) \cdot \hat{\mathbf{i}}_{i+1} \text{ and } \sin\hat{\theta}_i = (\hat{\mathbf{i}}_i \times \hat{\mathbf{i}}_{i+1}) \cdot \hat{\mathbf{k}}_i \tag{B.11}$$

Using Eqs. B.10 and B.11, the dual angles can be easily computed as

$$\begin{aligned} \hat{\alpha}_i &= \arctan 2(\sin\hat{\alpha}, \cos\hat{\alpha}) = \alpha_i + \varepsilon a_i \\ \hat{\theta}_i &= \arctan 2(\sin\hat{\theta}_i, \cos\hat{\theta}_i) = \theta_i + \varepsilon b_i \end{aligned} \tag{B.12}$$

Equation B.12 provides us with the values of the DH parameters, i.e., b_i, θ_i, a_i, and α_i.

B.4 Robot's Repeatability

As per ISO 9283 [4], five points within the robot's reach were considered for the repeatability experiments. The robot moved from one pose to another at 75% speed, i.e., 2 m/s. At each point, an image of the board of markers was captured and stored. For each point, the camera pose was calculated. This process was repeated 30 times and the repeatability of the robot was calculated using the expression provided in Appendix C. Figure B.4a shows the spatial location of the five points that are at the vertices and center of the cube of length 100 mm, whereas Fig. B.4b depicts the 30 scattered points for the pose corresponding to point 1.

Next, the parameters, x_j, y_j, and z_j, are the coordinates of the jth attained point. Then, the robot's repeatability in position, RP_l, is defined as

$$RP_l = \bar{l} + 3S_l \tag{B.13}$$

where $\bar{l} = \frac{1}{n}\sum_{j=1}^{n} l_j$ and $l_j = \sqrt{(x_j - \bar{x})^2 + (y_j - \bar{y})^2 + (z_j - \bar{z})^2}$. Additionally, $\bar{x} = \frac{1}{n}\sum_{j=1}^{n} x_j$, $\bar{y} = \frac{1}{n}\sum_{j=1}^{n} y_j$, and $\bar{z} = \frac{1}{n}\sum_{j=1}^{n} z_j$ which are the coordinates of the barycenter obtained after repeatedly coming to a point n times. Moreover, $S_l =$

---

[2] $\hat{\mathbf{k}}_1 \cdot \hat{\mathbf{k}}_2 = \mathbf{n}_1 \cdot \mathbf{n}_2 + \varepsilon(\mathbf{n}_1 \cdot \mathbf{k}_2^* + \mathbf{k}_1^* \cdot \mathbf{n}_2)$.

[3] $\hat{\mathbf{k}}_1 \times \hat{\mathbf{k}}_2 = \mathbf{n}_1 \times \mathbf{n}_2 + \varepsilon(\mathbf{n}_1 \times \mathbf{k}_2^* + \mathbf{k}_1^* \times \mathbf{n}_2)$.

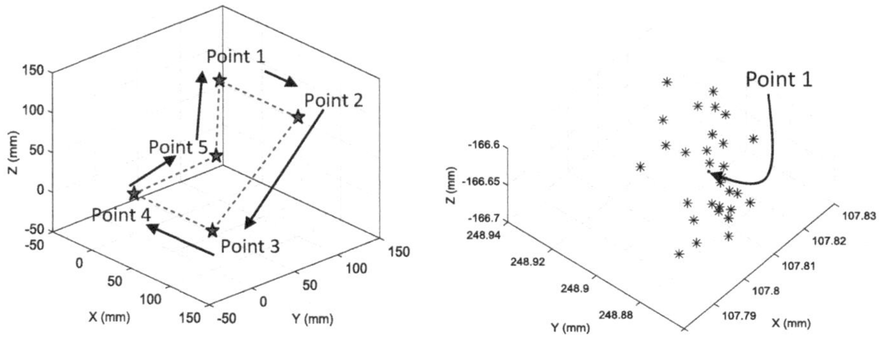

(a) Five points for repeatability measurement

(b) Scatter plot for Point 1

Fig. B.4 Scatterplot of points considered for repeatability measurement

$\sqrt{\frac{\sum_{j=1}^{n}(l_j-\bar{l})^2}{n-1}}$. Similarly, if A_j, B_j, and C_j be the roll, pitch, and yaw angles of the EE of the robot, respectively, corresponding to the pose at the jth point, the robot's repeatability for orientation with respect to each rotation is defined as

$$RR_A = \pm 3S_A = \pm 3\sqrt{\frac{\sum_{j=1}^{n}(A_j - \bar{A})}{n-1}} \tag{B.14}$$

$$RR_B = \pm 3S_B = \pm 3\sqrt{\frac{\sum_{j=1}^{n}(B_j - \bar{B})}{n-1}} \tag{B.15}$$

$$RR_C = \pm 3S_C = \pm 3\sqrt{\frac{\sum_{j=1}^{n}(C_j - \bar{C})}{n-1}} \tag{B.16}$$

where $\bar{A} = \frac{1}{n}\sum_{j=1}^{n} A_j$, $\bar{B} = \frac{1}{n}\sum_{j=1}^{n} B_j$, and $\bar{C} = \frac{1}{n}\sum_{j=1}^{n} C_j$. Additionally, $\bar{A}$, $\bar{B}$, and $\bar{C}$ are the mean values of the angles obtained after reaching the orientation n times.

B.5 Kinematic Chain Identification

For the identification of the kinematic chain, the motor rotation (MR), encoder counts (EC), and joint angles (JA) readings were taken from the controller which are listed in Table B.2.

Table B.2 Reading of motor rotation angles, encoder counts, and joint angles

	Joint 1			Joint 2			Joint 3		
S.N	MR	EC	JA	MR	EC	JA	MR	EC	JA
1	360	16419	2.89	360	16330	2.87	360	16427	4.15
2	720	32795	5.76	720	32723	5.75	720	32794	8.28
3	1080	49140	8.64	1080	49132	8.64	1080	49165	12.42
4	1440	65577	11.53	1440	65537	11.52	1440	65564	16.56
5	1800	81960	14.41	1800	81889	14.39	1800	81929	20.69
	Joint 4			Joint 5			Joint 6		
1	360	12280	4.83	360	12287	8.53	360	12283	14.92
2	720	24612	9.69	720	24570	17.05	720	24541	29.81
3	1080	36874	14.31	1080	36856	25.57	1080	36840	44.75
4	1440	49208	19.37	1440	49143	34.1	1440	49133	59.68
5	1800	61481	24.2	1800	61420	42.62	1800	61419	74.61

Table B.3 Encoder counts versus the joint angle

Joint space			Encoder counts		
θ_4	θ_5	θ_6	ε_4	ε_5	ε_6
10	10	10	25410	10433	8232
20	10	30	50821	13653	24355
30	10	20	76231	13274	15782
10	20	30	25408	28444	25038
40	50	60	101641	70542	49734
50	40	60	127052	55751	49052
60	50	40	152460	69784	32587

The coupling effect between the joints 4, 5, and 6 of KUKA KR5 Arc robot was estimated by relating the joint space and the encoder count values listed in Table B.3.

B.6 Inertia Dyad

The dyadic product takes two vectors and returns a second-order tensor. If **a** and **b** are two vectors expressed as $\mathbf{a} \equiv a_1\mathbf{i} + a_2\mathbf{j} + a_3\mathbf{k}$ and $\mathbf{b} \equiv b_1\mathbf{i} + b_2\mathbf{j} + b_3\mathbf{k}$, then the dyadic product of **a** and **b** is defined as

$$\begin{aligned}\mathbf{ab} =& a_1 b_1 \mathbf{ii}^T + a_1 b_2 \mathbf{ji}^T + a_1 b_3 \mathbf{ki}^T \\ &+ a_2 b_1 \mathbf{ij}^T + a_2 b_2 \mathbf{jj}^T + a_2 b_3 \mathbf{kj}^T \\ &+ a_3 b_1 \mathbf{ik}^T + a_3 b_2 \mathbf{jk}^T + a_3 b_3 \mathbf{kk}^T\end{aligned} \tag{B.17}$$

where $\mathbf{ii}^T$, $\mathbf{ij}^T$, $\cdots$, $\mathbf{kk}^T$ are known as unit dyads. A dyadic can contain a physical information such as inertia of an object. The inertia with nine elements is written as

$$\mathbf{I} \equiv \begin{bmatrix} I_{xx} & I_{xy} & I_{xz} \\ I_{yx} & I_{yy} & I_{yz} \\ I_{zx} & I_{zy} & I_{zz} \end{bmatrix} \tag{B.18}$$

The above expression can be written as sum of dyads scaled with moments of inertia or products of inertia, i.e.,

$$\mathbf{I} \equiv \begin{bmatrix} 1 & 0 & 0 \\ 0 & 0 & 0 \\ 0 & 0 & 0 \end{bmatrix} I_{xx} + \begin{bmatrix} 0 & 1 & 0 \\ 0 & 0 & 0 \\ 0 & 0 & 0 \end{bmatrix} I_{xy} + \cdots + \begin{bmatrix} 0 & 0 & 0 \\ 0 & 0 & 0 \\ 0 & 0 & 1 \end{bmatrix} I_{zz} \tag{B.19}$$

$$= \begin{bmatrix} \mathbf{ii}^T & (\mathbf{ij}^T + \mathbf{ji}^T) & (\mathbf{ik}^T + \mathbf{ki}^T) & \mathbf{jj}^T & (\mathbf{jk}^T + \mathbf{kj}^T) & \mathbf{kk}^T \end{bmatrix} \begin{bmatrix} I_{xx} \\ I_{xy} \\ I_{xz} \\ I_{yy} \\ I_{yz} \\ I_{zz} \end{bmatrix} \tag{B.20}$$

$$= \boldsymbol{\Omega} \bar{\mathbf{I}} \tag{B.21}$$

Note that Eq. (B.21) utilizes the symmetric property of the inertia tensor, i.e., $I_{xy} = I_{yx}$, $I_{xz} = I_{zx}$, and $I_{yz} = I_{zy}$. Then multiplication of the dyad with a vector can be expressed as

$$\mathbf{Ia} = \boldsymbol{\Omega} \bar{\mathbf{I}} \mathbf{a} \tag{B.22}$$

$$= \begin{bmatrix} \mathbf{ii}^T\mathbf{a} & (\mathbf{ij}^T + \mathbf{ji}^T)\mathbf{a} & (\mathbf{ik}^T + \mathbf{ki}^T)\mathbf{a} & \mathbf{jj}^T\mathbf{a} & (\mathbf{jk}^T + \mathbf{kj}^T)\mathbf{a} & \mathbf{kk}^T\mathbf{a} \end{bmatrix} \bar{\mathbf{I}} \tag{B.23}$$

$$= \boldsymbol{\Omega}(\mathbf{a}) \bar{\mathbf{I}} \tag{B.24}$$

Inertia tensor of a rigid body having volume v with respect to the link origin $\mathbf{I}_i$ is obtained using the parallel-axis theorem as

$$\mathbf{I}_i = \mathbf{I}_i^C - m_i \tilde{\mathbf{d}}_i \tilde{\mathbf{d}}_i \tag{B.25}$$

where I_i^C is the mass moment of inertia about the mass center C_i and $\mathbf{d}$ is the vector from link origin to the COM. Inertia tensor can be represented in the another frame, say $(i-1)^{st}$ frame as

$$[\mathbf{I}_o]_{i-1} = \mathbf{Q}_{i-1} [\mathbf{I}_o]_i \mathbf{Q}_{i-1}^T \tag{B.26}$$

B.7 Gauss–Jordan Elimination

To demonstrate a numerical method to check for the parameter dependence, the Gauss–Jordan elimination is presented. Let the nonlinear equation in six variables be taken as

$$f \equiv \sin\theta_1 x_1 + \cos\theta_2 x_2 + k_1 \sin\theta_3 x_3 + (\sin\theta_1 + 2\cos\theta_2)x_4 - \sin^2\theta_2 x_5 + 2\sin\theta_1 x_6 = 0 \tag{B.27}$$

where $\mathbf{x}_i$ $(i = 1, \cdots, 6)$ are the unknown variables with $k_1 = 0$. The above equation can be written in the variables separable form as

$$f = \underbrace{\left[\sin\theta_1 \;\; \cos\theta_2 \;\; k_1 \sin\theta_3 \;\; (\sin\theta_1 + 2\cos\theta_2) \;\; \sin^2\theta_2 \;\; 2\sin\theta_1\right]}_{\mathbf{Y}} \mathbf{x} \tag{B.28}$$

where $\mathbf{x} = [x_1, \cdots, x_6]^T$ is the six-dimensional vector of unknown. Let 10 randomly generated data for θ_1, θ_2, and θ_3 were input in Eq. (B.28) and stacked to give an overdetermined system. By performing the Gauss–Jordan elimination of the overdetermined matrix gives three non-zero rows and the rest all are zeros indicating rank of the system as 3 (Fig. B.5b). The new set of parameters, $\mathbf{x}_b \equiv [x_{b1}, x_{b2}, x_{b3}]$, was obtained as the combination of the variables x_1, x_2, x_4, and x_5. The variable x_3 does not affect the system and can be removed. Also the matrix $\mathbf{Y}$ gets changed accordingly by not considering the depended and zero column labeled as $\mathbf{Y}_b$ equivalent to $\mathbf{Y}_b \equiv [\sin\theta_1 \;\; \cos\theta_2 \;\; \sin^2\theta_2]$.

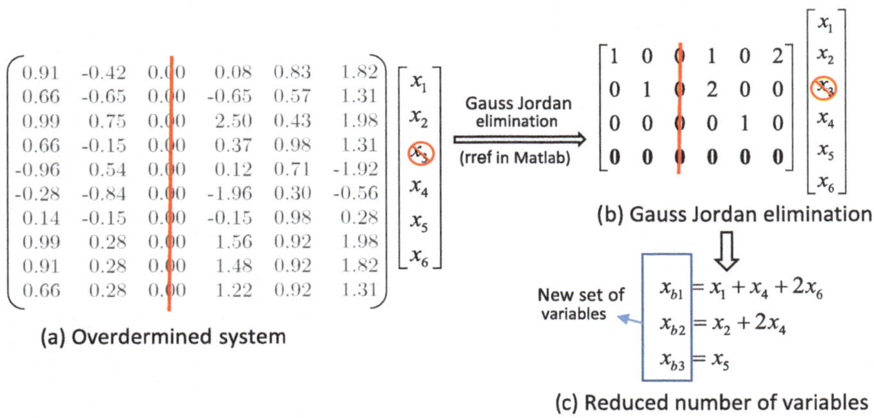

Fig. B.5 Gauss–Jordan elimination to obtain the reduced number of variables

References

1. Saha, S.K.: Introduction to Robotics, 2nd edn. Tata McGraw-Hill Education (2014b)
2. Denavit, J., Hartenberg, R.S.: A kinematic notation for lower-pair mechanisms based on matrices. Trans. ASME. J. Appl. Mech. **22**, 215–221
3. McCarthy, J.M.: Dual orthogonal matrices in manipulator kinematics. Int. J. Robot. Res. **5**(2), 45–51 (1986)
4. ISO/TR: Manipulating industrial robots—informative guide on test equipment and metrology methods of operation for robot performance evaluation in accordance with ISO 9283 (1995)

Appendix C
Code Snippets and Experimental Videos

In this appendix, several program snippets are provided to explain how the algorithms were built to implement the results reported in this monograph. The syntax of codes is subjected to vary as per the robot controller used and the version of open CV.

C.1 Peg-in-Tube Using KRL

The KRL is predecessor of C language. The programmer should take additional care while coding in KRL. The prerequisite for the development of code for peg-in-hole using KUKA Robot Language (KRL) is listed below:

- Understanding the safety procedures for operating the robot in a safe, reliable, and controlled manner.
- Coding in KRL (KUKA Robot Language) for understanding the motion depiction and control of the robot from the KRL environment.
- A working knowledge of Visual Studio and coding in C#.
- RSI (robot sensor interface) to comprehend the communication protocol between the PC running the C# code and the KRL environment in KUKA.

The process to teach the robot its location of the tool center point (TCP) with respect to the world coordinate system and assigning a coordinate system at TCP is defined as tool calibration. The point which is taught as tool which is depicted in Fig. C.1. The coordinate of this point is now known to the robot. The code snippet is shown in Fig. C.2 for the purpose of indicating the idea of data being send and received by the robot.

CAUTION: If the height of the peg is changed considerably (say $>$ 1 cm), then the recalibration of tool is required.

A. A. Hayat et al., *Vision Based Identification and Force Control of Industrial Robots*, Studies in Systems, Decision and Control 404,
https://doi.org/10.1007/978-981-16-6990-3

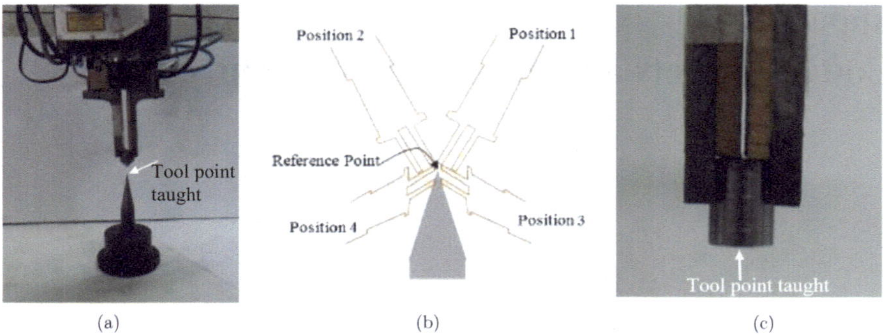

Fig. C.1 Calibration of tool for the peg-in-tube task

```xml
<ROOT> <CONFIG>
IP_NUMBER>192.168.1.60</IP_NUMBER> <!-- -->
PORT>6008</PORT> <!-- -->
PROTOCOL>TCP</PROTOCOL> <!-- -->
SENTYPE>ImFree</SENTYPE> <!-- -->
PROTOCOLLENGTH>Off</PROTOCOLLENGTH> <!-- -->
ONLYSEND>FALSE</ONLYSEND> <!-- -->
</CONFIG>
```

Defining configurations

```xml
<SEND> <ELEMENTS>
ELEMENT TAG="DEF_RIst" TYPE="DOUBLE" INDX="INTERNAL" UNIT="0" />
ELEMENT TAG="DEF_RSol" TYPE="DOUBLE" INDX="INTERNAL" UNIT="0" />
ELEMENT TAG="DEF_AIPos" TYPE="DOUBLE" INDX="INTERNAL" UNIT="0" />
ELEMENT TAG="DEF_ASPos" TYPE="DOUBLE" INDX="INTERNAL" UNIT="0" />
ELEMENT TAG="DEF_EIPos" TYPE="DOUBLE" INDX="INTERNAL" UNIT="0" />
ELEMENT TAG="DEF_ESPos" TYPE="DOUBLE" INDX="INTERNAL" UNIT="0" />
ELEMENT TAG="DEF_MACur" TYPE="DOUBLE" INDX="INTERNAL" UNIT="0" />
ELEMENT TAG="DEF_MECur" TYPE="DOUBLE" INDX="INTERNAL" UNIT="0" />
ELEMENT TAG="DEF_Delay" TYPE="LONG" INDX="INTERNAL" UNIT="0" />
ELEMENT TAG="DEF_Tech.C1" TYPE="DOUBLE" INDX="INTERNAL" UNIT="0" />
ELEMENT TAG="DiL" TYPE="LONG" INDX="1" UNIT="0" />
ELEMENT TAG="Digout.o1" TYPE="BOOL" INDX="2" UNIT="0" /> ELEMENT TAG="Digout.o2" TYPE="BOOL" INDX="3" UNIT="0" /> ELEMENT
TAG="Digout.o3" TYPE="BOOL" INDX="4" UNIT="0" /> ELEMENT TAG="ST_Source" TYPE="DOUBLE" INDX="5" UNIT="3601" />
</ELEMENTS> </SEND>
```

Defining send string by KUKA

```xml
<RECEIVE> ELEMENTS>
ELEMENT TAG="DEF_EStr" TYPE="STRING" INDX="INTERNAL" UNIT="0" />
ELEMENT TAG="RKorr.X" TYPE="DOUBLE" INDX="1" UNIT="1" HOLDON="1" />
ELEMENT TAG="RKorr.Y" TYPE="DOUBLE" INDX="2" UNIT="1" HOLDON="1" />
ELEMENT TAG="RKorr.Z" TYPE="DOUBLE" INDX="3" UNIT="1" HOLDON="1" />
ELEMENT TAG="RKorr.A" TYPE="DOUBLE" INDX="4" UNIT="0" HOLDON="1" />
ELEMENT TAG="RKorr.B" TYPE="DOUBLE" INDX="5" UNIT="0" HOLDON="1" />
ELEMENT TAG="RKorr.C" TYPE="DOUBLE" INDX="6" UNIT="0" HOLDON="1" />
ELEMENT TAG="AKorr.A1" TYPE="DOUBLE" INDX="7" UNIT="0" HOLDON="0" />
ELEMENT TAG="AKorr.A2" TYPE="DOUBLE" INDX="8" UNIT="0" HOLDON="0" />
ELEMENT TAG="AKorr.A3" TYPE="DOUBLE" INDX="9" UNIT="0" HOLDON="0" />
ELEMENT TAG="AKorr.A4" TYPE="DOUBLE" INDX="10" UNIT="0" HOLDON="0" />
ELEMENT TAG="AKorr.A5" TYPE="DOUBLE" INDX="11" UNIT="0" HOLDON="0" />
ELEMENT TAG="AKorr.A6" TYPE="DOUBLE" INDX="12" UNIT="0" HOLDON="0" />
<ELEMENT TAG="DEF_Tech.T2" TYPE="DOUBLE" INDX="INTERNAL" UNIT="0" />
ELEMENT TAG="DiO" TYPE="LONG" INDX="19" UNIT="0" HOLDON="1" />
</ELEMENTS> </RECEIVE> </ROOT>
```

Defining receive string by KUKA

```xml
<Rob Type="KUKA">
<AIPos A1=" 90.0 " A2="-90.00" A3=" 90.00 " A4=" 0 . 0 " A5=" 90.02 " A6=" 36.03 " />
<RI s t X="36.72" Y="807.74" Z=" 862.89 " A=" 126.03 " B="-0.02" C=" 179.99 "/>
<IPOC>4208995304</IPOC>
</Rob>
```

XML string received by the PC

```xml
<Sen Type=" F o r c eCt r l ">
<ESt r> ADU PC Connected </ ESt r>
<RKorr X=" 0 . 1 " Y="0" Z="0" A="0" B="0" C="0" />
<IPOC>4208995304</IPOC>
</Sen>
```

XML string sent by the PC

Fig. C.2 XML file for establishing the communication between PC and KUKA KR5 Arc

C.2 Hole Detection Using Vision

In order to insert the pellet in the randomly placed tube, vision comes into play. For detecting the hole, camera is oriented at a calibrated position above the vertical tube setup as shown in Chap. 5. After grabbing the image from the basler camera, image is processed using the Hough circles-based strategy and the center of holes are determined. User should be versed with

- OPEN CV camera calibration.
- C++ (Kuka server application).
- Different tools, i.e., Camera Frame, Suction Tool Frame, and Gripper Frame TRANSFORMATION should be known or user should find/identify as per the setup.
- Image processing (Fig. C.3).

C.3 Experimental Video files

The link to the folder containing the experimental video can be obtained by scanning the QR code in the end. The folder mainly contains the following files against the stated chapters:

- **Chapter** 2 **video**: The file named `Ch2_Bin_Pickup.mp4` demonstrates a) trial for pick-up of un-occluded pellets using only vision sensor and b) automated pick-up of pellets from the bin.
- **Chapter** 4 **video**: The file named `Ch4_Kin_Dyn_Idnetification.mp4` demonstrates (a) experimental setup for the monocular camera-based geometric parameters identification; (b) ROS-based interface and steps for identification of DH parameters using monocular camera; (c) inside axes 4, 5, and 6 of KUKA KR5 are used to determine the gear ratio; and (d) dynamic identification trajectory given go to KUKA KR5 Arc.
- **Chapters** 5 **and** 6 **videos**: The file named `Ch5_6_Control_Experiments.mp4` demonstrates (a) active/passive force control of industrial robot KUKA KR5 Arc; (b) Horizontal Pegintube task with tight tolerance between pellet and tube; and (c) Vertical Pegintube task with tight tolerance between pellet and tube.

Please scan above for videos and refer Appendix C.3 for files detail

```
void find_hole()
{
        kuka.startflag=11;
        cout<<"waiting   for   robot    to   reach   home    position   for Insertion"<<endl;
    while(kuka.calibflag!=11 && kuka.calibflag!=12)
        {//RaceCondition Sleep(1);}
        IplImage* grab_image; cout<<"Grabbinging Image"<<endl; grab_image = GrabImage();
    #if _DEBUG
        cout<<"Grab completed"<<endl; cout<<"Saving Image"<<endl; #endif
        std::stringstream imageNames;
        imageNames << "hole"<< imageNo << ".bmp"; std::string imageName = imageNames.str();
        char *image = const_cast<char*>(imageName.c_str());
        cvSaveImage(image,grab_image); Eigen::Vector3f homogeneous_coord; Eigen::Vector4f camPos;
        Eigen::Vector4f translation; Eigen::Vector3f pixel_coord; Eigen::Vector3f camera_coord; Eigen::Vector4f base_coord;
        Eigen::Matrix3f camera_internal;
        cv::Mat distCoeffs = (Mat1d(1, 5) << -0.201788, 0, 0, 0, 0);
        Eigen::Vector4f camera_homo_coord; Eigen::Matrix4f T_T_C; Eigen::Matrix4f B_T_T; Eigen::Matrix4f projectionMat;
        cv::Mat rgbTubes,disttubes;
        camera_internal<<      2328.556745,    0.000000,    1254.058997, 0.000000, 2334.311366, 1012.072590, 0.000000, 0.000000, 1.000000;
        Mat cameraMatrix = (Mat1d(3, 3) << 2328.556745, 0.000000, 1254.058997,  0.000000,    2334.311366, 1012.072590, 0.000000,    0.000000, 1.000000);;
        rgbTubes = cv::imread(imageName, 1 );
        cvtColor(rgbTubes, disttubes, CV_BGR2GRAY); undistort(disttubes, tubes,     cameraMatrix,    distCoeffs, cameraMatrix );
        HoughCircles( tubes, holes, CV_HOUGH_GRADIENT, 1.5, 700, 130, 55, 75, 95 );

        Point center(cvRound(holes[0][0]), cvRound(holes[0][1]));
        int radius = cvRound(holes[0][2]);
        circle( tubes, center, 3, Scalar(255,0,0), -1, 8, 0 );
        circle( tubes, center, radius, Scalar(255,0,0), 3, 8, 0 ); // draw the circle outline//

        imwrite("hough.jpg",tubes);
        cout<<center<<endl;
        //Internal Parameters pixel_coord<<center.x,center.y,1;
        homogeneous_coord=camera_internal.inverse() *pixel_coord;
        /*center.x = (center.x - c_x) / f_x; center.y = (center.y - c_y) / f_y;*/
        cout<<"homg coord."<<homogeneous_coord<<endl;
        //Back Projection camPos<<kuka.x_k,kuka.y_k,kuka.z_k,1; cout<<"tubePose"<<camPos<<endl;

        translation<< 11.039197, 65.119311, 133.896502, 0;

        camPos = translation + camPos;
        cout<<"tubePose"<<camPos<<endl;
        //tubeHeight = tubePos[2] + kuka.z_k; tubeHeight = camPos[2]+ tubeHeight;
        camera_coord=homogeneous_coord*tubeHeight;

        cout<<"camera coord."<<camera_coord<<endl;
        /*insertX = center.x * tubeHeight; insertY = center.y * tubeHeight;
        insertZ = tubeHeight;*/
            camera_homo_coord[0]=camera_coord[0];
            camera_homo_coord[1]=camera_coord[1];
            camera_homo_coord[2]=camera_coord[2]; camera_homo_coord[3]=1;

        cout<<"homg homg. coord."<<camera_homo_coord<<endl;
        //Camera to tool

    float x = kuka.x_k,y=kuka.y_k,z=kuka.z_k,a=kuka.a_k* PI / 180.0,b=kuka.b_k* PI / 180.0,c=kuka.c_k* PI /
    180.0;
              B_T_T << cos(a)*cos(b), cos(a)*sin(b)*sin(c) - cos(c)*sin(a),
    sin(a)*sin(c) + cos(a)*cos(c)*sin(b), x ,
                            cos(b)*sin(a), cos(a)*cos(c) + sin(a)*sin(b)*sin(c), cos(c)*sin(a)*sin(b) -
    cos(a)*sin(c), y ,
                                -sin(b),
    cos(b)*sin(c),         cos(b)*cos(c), z ,0,0,0, 1;
    cout<<"B_T_T"<< B_T_T<<endl;

            T_T_C << -0.605277, 0.794761, 0.044664, 11.039197,
                     -0.795100, -0.606319, 0.013937, 65.119311,
                     0.038158, -0.027077, 0.998905, 133.896502,
                     0,          0,         0,
        1;
            cout<<"T_T_C "<<T_T_C<<endl;


            cout<< "trasansform"<< B_T_T*T_T_C.inverse()<<endl; base_coord=B_T_T * T_T_C.inverse()

            *camera_homo_coord; cout<< B_T_T<<endl;
            insertX=base_coord[0]; insertY=base_coord[1]; insertZ=base_coord[2];

             cout<< base_coord<<endl;

             // 300 / 130 /113

            /*[X,Y,Z] =  [insertX, inserty,insertZ]; base_coord=camera_coord**/
            while(kuka.calibflag!=12)
            {//RaceCondition
                Sleep(1);
            }
            kuka.startflag=0;

            cout<<"Found Hole"<<endl;
            /*namedWindow( " hole", WINDOW_NORMAL | CV_WINDOW_KEEPRATIO
    );
            imshow( " hole", tubes); waitKey(10); cvDestroyAllWindows();*/
    }
```

Fig. C.3 Code snippet of vision-based hole detection using OpenCV

The manufacturer's authorised representative in the EU is Springer Nature Customer Service Centre GmbH, Europaplatz 3, 69115 Heidelberg, Germany. If you have any concerns regarding our products, please contact ProductSafety@springernature.com

Printed and bound by CPI Group (UK) Ltd, Croydon, CR0 4YY
06/10/2026
02225252-0004